Solving Problems in Fluid Mechanics

Volume 1

Solving Problems in Fluid Mechanics
Volume 1

J. F. Douglas MSc PhD DIC ACGI CEng MICE MIMechE MIStructE

Copublished in the United States with
John Wiley & Sons, Inc., New York

Longman Scientific & Technical
Longman Group UK Limited
Longman House, Burnt Mill, Harlow,
Essex CM20 2JE, England
and Associated Companies throughout the world.

Copublished in the United States with
John Wiley & Sons Inc., 605 Third Avenue, New York, NY 10158.

First published by Pitman Publishing Limited in 1970 under the title
Solution of Problems in Fluid Mechanics Part I
All metric edition first published 1975
Ninth impression 1984
This edition published by Longman Scientific & Technical in 1986 under the title
Solving Problems in Fluid Mechanics Volume I
Second impresssion 1987
Third impression 1987
Fourth impression 1988
Fifth impression 1989
Sixth impression 1991
Seventh impression 1992
Eighth impression 1994

British Library Cataloguing in Publication Data
Douglas, J. F.
  [Solution of problems in fluid mechanics]
  Solving problems in fluid mechanics.—All-metric
  ed. - (Solving problems)
  Vol. 1
  1. Fluid mechanics
  I. [Solution of problems in fluid mechanics]
  II. Title
  620.1′06   TA357

ISBN 0-582-28641-7
ISBN 0-470-20775-2 (USA only)

Produced by Longman Singapore Publishers (Pte) Ltd.
Printed in Singapore.

Contents

Contents

Preface

Some students find it difficult to learn from an ordinary textbook because they tend to read it as one might read a novel and fail either to appreciate what is being set out in each section or to study each mathematical step in detail. The result is that the student ends his reading with a glorious feeling of knowing it all and with, in fact, no understanding of the subject whatsoever. To avoid this undesirable end I have adopted for this book an older but traditional presentation of the subject in the form of a catechism with question and answer.

Even if the reader does achieve an understanding of the principles of the subject, there still remains that familiar gap between knowing and actually doing that everyone knows so well. It is one thing to know how a job should be done but it needs experience to do it with confidence. I have therefore provided for each chapter exercises with answers giving an opportunity for plenty of practice before the readers go off to solve their own problems.

This is definitely a textbook, not a book of worked examples, and in its pages the reader will find all the definitions and theory needed presented in question and answer form with selected problems fully worked out and suitable exercise questions to test the understanding of each section.

The material included in this volume covers the elementary work in Fluid Mechanics for engineering students in Universities, Polytechnics and Colleges of Higher Education. More advanced work is covered in the subsequent volume.

In writing for a wide readership it is impossible to satisfy everyone. For instance in the simple matter of density, some swear that mass density is the only true density and that all else is an abomination, others find specific weight a useful concept, yet others say that specific gravity should be abolished. Again while SI units are preferred by many, other systems are still in use and perhaps merit a reference. While I may prefer mass density and am committed to SI units, I have tried to meet the requirements of others where possible.

It has been suggested that some attention should be given to the use of computers in the solving of problems in fluid mechanics. For those who want them, examples will be found in *Fluid Mechanics* by Douglas, Gasiorek and Swaffield (Pitman, 2nd Edn. 1985). I have preferred to concentrate here on the principles of fluid mechanics which do not change and on the derivation of the corresponding algebraic equations. How these equations are solved will depend on the facilities available – computer, calculator, abacus, slide rule, mathematical tables and pencil and paper. In engineering, success depends on the reliability of the results achieved, not on the method of achieving them.

I would like to express my appreciation of the assistance which I have received from my former colleagues in the teaching profession. I am particularly indebted to Dr. R. D. Matthews for his advice on the preparation of this new text and for the provision of examples and exercises with particular reference to Chapter 14.

When author, printer and publisher have all done their best, some errors may still remain. For these I apologise and I will be glad to receive any correction or constructive criticism.

John Douglas   August 1985

Introduction

FLUID MECHANICS is the section of Applied Mechanics concerned with the statics and dynamics of liquids and gases. The same ideas of momentum and energy, etc. used in ordinary mechanics can be applied, but frequently fluid mechanics is concerned with streams of fluid instead of individual bodies or particles.

Hydraulics (from the Greek word for water) is the study of the problems of the flow and storage of water but is often applied to other liquids, as for example in the case of "hydraulic" control gear usually using oil as the operating fluid.

A fluid can offer no permanent resistance to any force causing change of shape. Fluids flow under their own weight and take the shape of any solid body with which they are in contact.

Change of shape is caused by shearing forces; therefore if shearing forces are acting in a fluid it will flow. Conversely, if a fluid is at rest there can be no shearing forces in it, and all forces are perpendicular (normal) to the planes on which they act.

Fluids are divided into liquids and gases. A *liquid* is difficult to compress; a given mass occupies a fixed volume irrespective of the size of the container holding it and a "free surface" is formed as a boundary between the liquid and the air above it. A *gas* is easily compressed; it expands to fill any vessel in which it is contained and it does not form a free surface.

Distinctions between a solid and a fluid are: (1), up to the limit of elasticity the deformation of a solid is such that the strain is proportional to the applied stress; for a fluid the rate of strain is proportional to the stress; (2), the strain of a solid is independent of the time of application of the force and if the elastic limit is not exceeded the deformation disappears when the stress is removed, but a fluid continues to flow as long as the stress is applied and does not recover its original form when the stress is removed.

Units and dimensions

SI Units

The United Kingdom has now adopted the system of metric units known as the Système International d'Unités, abbreviated to SI. In due course these will replace the old British units such as the pound, poundal and foot as the only legal system of measurement.

The Système International (SI) has six basic units which are arbitrarily defined. These are:

length: metre (m)
mass: kilogramme (kg)
time: second (s or sec)
electric current: ampere (A)
absolute temperature: kelvin (K)
luminous intensity: candela (cd)

All other units are derived from these fundamental units, since the SI is a coherent system in which the product or quotient of any two unit quantities within the system is the unit of the resultant quantity. For example, the unit of velocity is obtained by dividing the unit of distance, the metre, by the unit of time, the second, and will therefore be metres per second. The relation between mass and force is established by making the constant of proportionality in Newton's second law equal to unity so that

$$\text{Force} = \text{mass} \times \text{acceleration}$$

The unit of force will be the product of the unit of mass (kilogramme) and the unit of acceleration (metre/sec^2) which is the kilogramme-metre/sec^2 and is known as the newton.

Larger or smaller units are formed by adding a prefix to the basic unit. For example one thousandth part of a metre is a *millimetre* while one thousand metres is a *kilometre*. The following prefixes are currently in use:

Multiples			*Sub-multiples*		
deca	da	$= \times 10$	*deci*	d	$= \times 10^{-1}$
hecto	h	$= \times 10^2$	*centi*	c	$= \times 10^{-2}$
kilo	k	$= \times 10^3$	milli	m	$= \times 10^{-3}$
mega	M	$= \times 10^6$	micro	μ	$= \times 10^{-6}$
giga	G	$= \times 10^9$	nano	n	$= \times 10^{-9}$
tera	T	$= \times 10^{12}$	pico	p	$= \times 10^{-12}$
			femto	f	$= \times 10^{-15}$
			atto	a	$= \times 10^{-18}$

The prefixes shown in italics are not preferred but are in common use. Another unit in common use is the metric tonne $= 10^3$ kg $= 2205$ lb.

Other systems of units

While SI units are the preferred system, there are other systems which remain of importance in various parts of the world and in particular fields of activity. The foot-pound-second system (fps), the centimetre-gramme-second system (cgs) and the metre-kilogramme-second system (MKS) have been used extensively and no doubt will continue to be used. They are coherent systems based on a constant of unity in Newton's second law and appear in two forms: in the absolute systems the unit of mass is a fundamental unit and the unit of force is derived, whereas in the technical system the unit of force is a fundamental unit and the unit of mass is derived using Newton's second law (Table I).

The MKS absolute units, so far as mechanics is concerned, correspond with SI units and it seems possible that MKS technical units may continue in use for sometime alongside SI units.

In solving problems it is essential to keep to one system of units only. If the data are in different systems they should be converted immediately to the system selected.

Table I

Quantity	fps		cgs		MKS
	Absolute	Technical	Absolute	Technical	Technical
Length	ft	ft	cm	cm	m
Time	sec	sec	sec	sec	sec
Mass	lb-mass	slug	g-mass	981 g	9·81 kg
Force or weight	poundal	lb-force	dyne	g-force	kg-force

1 slug = 32·2 lb-mass, 1 g-force = 981 dynes, 1 lb-force = 32·2 poundals.

Dimensions

The units chosen for measurement do not affect the quantity measured. One kilogramme of water means exactly the same as 2·2046 lb of water. It is sometimes convenient not to use any particular system but to think in terms of mass, length, time, force, temperature, etc.

In mechanics all quantities can be expressed in terms of the fundamental dimension of mass M, length L and time T.

Thus
$$acceleration = \frac{distance}{(time)^2}$$

so that

$$dimensions\ of\ acceleration = \frac{dimension\ of\ distance}{(dimension\ of\ time)^2} = \frac{L}{T^2}$$

Similarly
$$Force = mass \times acceleration$$

so that

Dimension of Force = dimension of mass × dimension of acceleration

$$= \frac{ML}{T^2}$$

The dimensions and SI units of common quantities are shown in Table II.

Dimensional equations

If an equation is to represent something which is physically real the terms on both sides must be of the same sort (for example, all forces) as well as the two sides being numerically equal, otherwise the equation is meaningless. Every term must have the same dimensions so that like is compared with like.

Table II Dimensions and units of common quantities

Quantity	Defining equation	Dimensions	Unit	Symbol
Geometrical				
Angle	Arc/radius (a ratio)	$[M^0L^0T^0]$	radian	rad
Length	(Including all linear measurement)	$[L]$	metre	m
Area	Length × Length	$[L^2]$	square metre	m²
Volume	Area × Length	$[L^3]$	cubic metre	m³
First moment of area	Area × Length	$[L^3]$	metre cubed	m³
Second moment of area	Area × Length²	$[L^4]$	metre to fourth power	m⁴
Strain	Extension/Length	$[L^0]$	a ratio	
Kinematic				
Time		$[T]$	second	s
Velocity, linear	Distance/Time	$[LT^{-1}]$	metre per second	ms⁻¹
Acceleration, linear	Linear velocity/Time	$[LT^{-2}]$	metre per second squared	ms⁻²
Velocity, angular	Angle/Time	$[T^{-1}]$	radians per second	rad s⁻¹
Acceleration, angular	Angular velocity/Time	$[T^{-2}]$	radians per second squared	rad s⁻²
Volume rate of discharge	Volume/Time	$[L^3T^{-1}]$	cubic metres per second	m³s⁻¹
Dynamic				
Mass	Force/Acceleration	$[M]$	kilogramme	kg
Force	Mass × Acceleration	$[MLT^{-2}]$	newton = kilogramme-metre per second²	N = kgms⁻²
Weight	Force	$[MLT^{-2}]$	newton	N
Mass density	Mass/Volume	$[ML^{-3}]$	kilogrammes per cubic metre	kg m⁻³
Specific weight	Weight/Volume	$[ML^{-2}T^{-2}]$	newtons per cubic metre	N m⁻³
Specific gravity	Density/Density of water	$[M^0L^0T^0]$	a ratio	–
Pressure (intensity)	Force/Area	$[ML^{-1}T^{-2}]$	newton per square metre = pascal	Nm⁻² = Pa
Stress	Force/Area	$[ML^{-1}T^{-2}]$	newton per square metre	Nm⁻²
Elastic modulus	Stress/Strain	$[ML^{-1}T^{-2}]$	newton per square metre	Nm⁻²
Impulse	Force × Time	$[MLT^{-1}]$	newton seconds	N s
Mass moment of inertia	Mass × Length²	$[ML^2]$	kilogramme-metre squared	kg m²
Momentum, linear	Mass × Linear velocity	$[MLT^{-1}]$	kilogramme-metre per second	kg m s⁻¹
Momentum, angular	Moment of inertia × Angular velocity	$[ML^2T^{-1}]$	kilogramme-metre squared per second	kg m²s⁻¹
Work, energy	Force × Distance	$[ML^2T^{-2}]$	newton-metre = joule	Nm = J
Power	Work/Time	$[ML^2T^{-3}]$	joule per second = watt	JS⁻¹ = W
Moment of a force	Force × Distance	$[ML^2T^{-2}]$	newton-metre	Nm
Viscosity, dynamic	Shear stress/Velocity gradient	$[ML^{-1}T^{-1}]$	kilogrammes per metre-second (= 10 poise)	kg m⁻¹s⁻¹
Viscosity, kinematic	Dynamic viscosity/Mass density	$[L^2T^{-1}]$	metre squared per second	m²s⁻¹
Surface tension	Energy/Area	$[MT^{-2}]$	newton per metre = kilogrammes per second squared	Nm⁻¹ = kgs⁻²

Example. The equation $v^2 = u^2 + 2as$ gives the final velocity v of a body which started with an initial velocity u and received an acceleration a for a distance s. When dimensions are substituted for the quantities each term must have the same dimensions if the equation is true.

The dimensions of the quantities are $v = LT^{-1}, u = LT^{-1}, a = LT^{-2}, s = L$

$$\text{Dimensions of } v^2 \text{ are} \quad (LT^{-1})^2 = L^2T^{-2}$$

$$\text{Dimensions of } u^2 \text{ are} \quad (LT^{-1})^2 = L^2T^{-2}$$

$$\text{Dimensions of } 2as \text{ are} \quad (LT^{-2}) \times L = L^2T^{-2}$$

All three terms have the same dimensions and the equation is dimensionally correct and could represent a real event.

A check on dimensions will not show whether any pure numbers in the equation are correct since pure numbers are ratios and have the dimension of unity.

Note that some practical formulae used by engineers do not appear to be dimensionally correct. For example, a formula for the volume in m^3/s per second Q flowing over a rectangular weir of width B metres when the depth over the sill is H metres is $Q = 1.79BH^{3/2}$. The dimensions of the left-hand side (m^3/s) are L^3T^{-1}. The dimensions of the right-hand side are apparently $L^{5/2}$. The reason for the difference is that the coefficient 1.79 is not a pure number but is the numerical value in SI units of $0.57\sqrt{g}$ which has dimensions $L^{1/2}T^{-1}$, giving dimensional agreement between the two sides.

Use of dimensions for finding conversion factors

This is shown in the following example.

Example. The coefficient of dynamic viscosity μ of water at 95°F is 1.505×10^{-5} ft slug sec units; what is the value in (a) poises, (b) SI units. From Table II the dimensions of μ are M/LT.

$(a) \quad \therefore \quad \dfrac{\mu_{(poises)}}{\mu_{(ft\ slug\ sec)}} = \dfrac{(M/LT)\ \text{in cgs absolute units}}{(M/LT)\ \text{in ft slug sec units}}$

$= \dfrac{\text{Mass (cgs abs)}}{\text{Mass (ft slug sec)}} \times \dfrac{\text{length (ft slug sec)}}{\text{length (cgs)}} \times \dfrac{\text{time (ft slug sec)}}{\text{time (cgs)}}$

$$1\ \text{slug} = 32.2\ \text{lb mass} = 32.2 \times 453.6\ \text{g mass}$$

$$1\ \text{ft} = 30.48\ \text{cm}$$

The time unit is 1 sec in both systems.

$$\therefore \quad \mu_{(poises)} = \mu_{(ft\ slug\ sec)}\ \frac{32.2 \times 453.6}{1} \times \frac{1}{30.48} \times \frac{1}{1}$$

$$= \frac{1.505 \times 10^{-5} \times 32.2 \times 453.6}{30.48} = 7.2 \times 10^{-3}\ \text{poises}$$

$(b) \quad \dfrac{\mu_{(SI)}}{\mu_{(ft\ slug\ sec)}} = \dfrac{(M/LT)\ \text{in ft slug sec units}}{(M/LT)\ \text{in SI units}}$

$= \dfrac{\text{Mass (SI units)}}{\text{Mass (ft slug sec)}} \times \dfrac{\text{length (ft slug sec)}}{\text{length (SI units)}} \times \dfrac{\text{time (ft slug sec)}}{\text{time (SI units)}}$

$$1\ \text{slug} = 32.2\ \text{lb-mass} = 32.2 \times 0.4536\ \text{kg-mass}$$

$$1\ \text{ft} = 0.3048\ \text{m}$$

The time unit is 1 sec in both systems.

$$\mu_{(SI\ units)} = \mu_{(ft\ slug\ sec)}\ \frac{32.2 \times 0.4536}{1} \times \frac{1}{0.3048} \times \frac{1}{1}$$

$$= \frac{1.505 \times 10^{-5} \times 32.2 \times 0.4536}{0.3048} = 7.2 \times 10^{-4}\ \text{kg/m-s}$$

Properties of fluids

Density

There are three forms of density which must be carefully distinguished.

1. *Mass density* ρ (Gk., rho) is the mass per unit volume. SI unit, kg/m^3 (fps absolute unit, lb-mass/ft^3; technical unit, slug-mass/ft^3).

2. *Specific weight* w is the weight per unit volume. SI unit, N/m^3 (fps absolute unit, poundal/ft^3; technical unit, lb-wt/ft^3).

Since weight = mass × gravitational acceleration

$$w = \rho g$$

3. *Specific gravity*, or relative density s, is the ratio of the weight of a substance to the weight of an equal volume of water at 4°C,

$$s = \frac{w \text{ for substance}}{w \text{ for water}} = \frac{\rho \text{ for substance}}{\rho \text{ for water}}$$

Viscosity

A fluid at rest cannot resist shearing forces but once it is in motion shearing forces are set up between layers of fluid moving at different velocities. The viscosity of the fluid determines its ability to resist these shearing stresses (see Chap. 13).

The *Coefficient of Dynamic Viscosity* μ (Gk., mu) is defined as the shear force per unit area required to drag one layer of fluid with unit velocity past another layer unit distance away from it in the fluid. SI unit, $N\text{-}s/m^2$ or kg/m-s.

(fps absolute unit is lb-mass/ft-sec. Technical unit, slug/ft-sec.)

In the absolute cgs system of units the unit of viscosity is the poise which is divided into 100 centipoises. (1 poise = 1 g/cm-s.)

Kinematic Viscosity v (Gk., nu) is the ratio of dynamic viscosity to mass density

$$v = \frac{\mu}{\rho}$$

Note that, if μ is in kg/m-s, ρ must be in kg/m^3, thus the units of v are independent of mass. The SI unit is m^2/s. (fps unit, ft^2/s.) In the cgs system the unit is the stoke which is divided into 100 centistokes.

Variation of Viscosity with Temperature. The viscosity μ of liquids decreases with increase of temperature, but the viscosity of gases increases with increase of temperature.

Poiseuille showed that

$$\mu = \mu_0 \left(\frac{1}{1 + at + bt^2} \right)$$

where μ = coefficient of viscosity at t°C, μ_0 = coefficient of viscosity at 0°C, a and b are constants.

For water μ_0 = 0.0179 poise = 0.00179 kg/m-s, a = 0.033368 and b = 0.000221.

Surface tension
σ (Gk., sigma)

Within the body of a liquid a molecule is attracted equally in all directions by the other molecules surrounding it, but at the surface between liquid and air the upward and downward attractions are unbalanced. The liquid surface behaves as if it were an elastic membrane under tension. This surface tension is the same at every point on the surface and acts in the plane of the surface normal to any line in the surface. Surface tension is not affected by the curvature of the surface, and it is constant at a given temperature for the surface of separation of two particular substances. Increase of temperature causes a decrease of surface tension.

Surface tension causes drops of liquid to tend to take a spherical shape and is also responsible for capillary action which causes a liquid

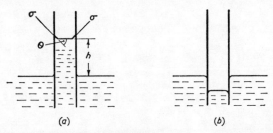

(a)   (b)

Figure 1

to rise in a fine tube when its lower end is inverted in a liquid which wets the tube (Fig. 1a). If the liquid does not wet the tube it will be depressed in the fine tube below the surface outside.

If θ is the angle of contact between liquid and solid, upward pull due to surface tension $= \sigma \pi d \cos \theta$ where d = diameter of tube.

Putting h = height liquid is raised and w = sp. wt of liquid

$$\text{weight of liquid raised} = w \frac{\pi}{4} d^2 h$$

so that

$$\sigma \pi d \cos \theta = w \frac{\pi}{4} d^2 h$$

$$h = \frac{4\sigma \cos \theta}{wd}$$

Capillary action is a source of error in reading gauge glasses. For water in a tube 6mm in diameter h will be 4·5mm, while for mercury the corresponding figure is $-1\cdot5$mm.

Compressibility

For liquids the relationship between change of pressure and change of volume is given by the bulk modulus K.

$$\text{Bulk modulus} = \frac{\text{change in pressure intensity}}{\text{volumetric strain}}$$

$$= \frac{\text{change in pressure intensity}}{(\text{change in volume/original volume})}$$

The relation between pressure and volume for a gas can be found from the gas laws.

For all perfect gases $pv = RT$, where p = absolute pressure, v = specific volume = $1/w = 1/\rho g$, T = absolute temperature, R = gas constant.

If changes occur isothermally (at constant temperature) pv = constant.

If changes occur adiabatically (without gain or loss of heat), pv^γ = constant, where γ = ratio of specific heat at constant pressure to specific heat at constant volume.

1

Static pressure and head

A force or pressure is exerted by a fluid on the surfaces with which it is in contact, or by one part of a fluid on the adjoining part. The *intensity of pressure* at any point is the force exerted on unit area at that point and is measured in newtons per square metre (*pascals*) in SI units (pounds per square foot in fps technical units). An alternative metric unit is the bar, which is 10^5 N/m². In practice intensity of pressure is abbreviated to *pressure*.

1.1 Pressure intensity

A mass m of 50 kg acts on a piston of area A of 100 cm². What is the intensity of pressure on the water in contact with the underside of the piston if the piston is in equilibrium.

Solution.

$$\text{Force acting on piston} = mg$$

$$= 50 \times 9{\cdot}81 = 490{\cdot}5\,\text{N}$$

$$\text{Area of piston } A = 100\,\text{cm}^2 = \frac{1}{100}\,\text{m}^2$$

$$\text{Intensity of pressure} = \frac{\text{Force}}{\text{Area}} = \frac{490{\cdot}5}{0{\cdot}01}\,\text{N/m}^2$$

$$= 4{\cdot}905 \times 10^4\,\text{N/m}^2$$

1.2 Pressure and depth

Find the intensity of pressure p at a depth h below the surface of a liquid of specific weight $w = \rho g$ if the pressure at the free surface is zero.

A diver is working at a depth of 18 m below the surface of the sea. How much greater is the pressure intensity at this depth than at the surface? Specific weight of sea water is 10 000 N/m³.

Solution. The column of liquid (Fig. 1.1) of cross-sectional area A extending vertically from the free surface to the depth h is in equilibrium in the surrounding liquid under the action of its weight acting downwards, the pressure force on the bottom of the column acting upwards, and the forces on the sides due to the surrounding liquid which must act horizontally since there can be no tangential (shearing) forces in a liquid at rest. For vertical equilibrium,

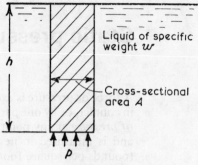

Figure 1.1

Force exerted on base = Weight of column of liquid

Intensity of pressure × area of base

$$= \text{Weight per unit volume} \times \text{volume of column}$$

$$pA = w \cdot Ah$$

$$p = wh = \rho gh \quad \text{since } w = \rho g$$

Since the same relation applies wherever the column is taken, it follows that

The intensity of pressure is the same at all points in the same horizontal plane in a liquid at rest.

Putting $w = 10000\,\text{N/m}^3$ and $h = 18\,\text{m}$

$$p = 10000 \times 18$$

$$= 180000\,\text{N/m}^2$$

1.3 Pressures at a point

Show that the intensity of pressure at a point in a fluid at rest is the same in all directions.

Solution. In Fig. 1.2, p_1 is the intensity of pressure on the horizontal face AB of a very small prism ABC of width s surrounding the given point, p_2 is the intensity of pressure on the vertical face BC and p_3 is the intensity of pressure on the face AC inclined at any angle θ to the horizontal.

$$\text{Force on face AB} = p_1 \times \text{AB} \times s$$
$$\text{Force on face BC} = p_2 \times \text{BC} \times s$$
$$\text{Force on face AC} = p_3 \times \text{AC} \times s$$

If the fluid is at rest, these forces are in equilibrium and are perpendicular to the faces on which they act.

Resolving vertically,

$$p_1 \times AB \times s = p_3 \times AC \times s \times \cos \theta$$

but by trigonometry

$$AC \cos \theta = AB$$

and so

$$p_1 = p_3$$

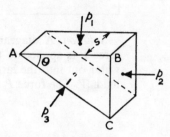

Figure 1.2

Resolving horizontally,

$$p_2 \times BC \times s = p_3 \times AC \times s \times \sin \theta$$

but

$$AC \sin \theta = BC$$

and so

$$p_2 = p_3$$

Therefore $p_1 = p_2 = p_3$ and since AC is at any angle θ to AB, p_3 is the intensity of pressure in any direction. Thus the intensity of pressure at a point is the same in all directions in a fluid at rest.

Pressure head

The pressure p at a point in a fluid can be expressed in terms of the height h of the column of the fluid which causes the pressure, or which would cause an equal pressure if the actual pressure is applied by other means. From Example 1.2, $p = wh = \rho g$ and the height h is called the *pressure head* at the point. It is measured as a length (e.g. in metres) of fluid. The name of the fluid must be given.

1.4   Pressure and head

> Find the head h of water corresponding to an intensity of pressure p of 340 000 N/m^2. The mass density ρ of water is 10^3 kg/m^3.

Solution. Since $p = \rho g h$

$$\text{Head of water } h = \frac{p}{\rho g} = \frac{340\ 000}{10^3 \times 9 \cdot 81} = \textbf{34.7 m}$$

1.5 Hydraulic jack

Explain with a diagram the action of an hydraulic jack. A force P of 850 N is applied to the smaller cylinder of an hydraulic jack. The area a of the small piston is 15 cm² and the area A of the larger piston is 150 cm². What load W can be lifted on the larger piston (a) if the pistons are at the same level, (b) if the large piston is 0.75 m below the smaller? The mass density ρ of the liquid in the jack is 10^3 kg/m³.

Solution. A diagram of an hydraulic jack is shown in Fig. 1.3. A force P is applied to the piston of the small cylinder and forces oil or water out into the large cylinder thus raising the piston supporting the load W. The force P acting on area a produces a pressure p_1 which is

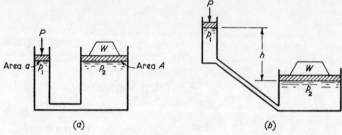

Figure 1.3

transmitted equally in all directions through the liquid (see 1.3). If the two pistons are at the same level, the pressure p_2 acting on the larger piston must equal p_1.

$$\text{Now } p_1 = \frac{P}{a} \quad \text{and} \quad p_2 = \frac{W}{A}$$

$$\text{If } p_1 = p_2, \qquad \frac{P}{a} = \frac{W}{A} \quad \text{or} \quad P = W\frac{a}{A}$$

Thus a small force P can raise a larger load W. The jack has a mechanical advantage of A/a.

(a) Putting $P = 850\,\text{N}$, $a = 15/10000\,\text{m}^2$, $A = 150/10000\,\text{m}^2$

$$\frac{P}{a} = \frac{W}{A}$$

so that
$$W = P\frac{A}{a} = 850 \times \frac{1 \cdot 5}{0 \cdot 15} = 8\,500\,\text{N}$$

$$\text{Mass lifted} = \frac{W}{g} = \frac{8\,500}{9 \cdot 81} = \textbf{868\,kg}$$

(b) If the larger piston is a distance h below the smaller, the pressure p_2 will be greater than p_1, due to the head h, by an amount ρg, where ρ is the mass density of the liquid.

$$p_2 = p_1 + \rho g h$$

Putting $p_1 = P/a = 850/15 \times 10^{-4} = 56{\cdot}7 \times 10^4\,\text{N/m}^2$, $\rho = 10^3\,\text{kg/m}^3$ and $h = 0{\cdot}75\,\text{m}$

$$p_2 = 56{\cdot}7 \times 10^4 + (10^3 \times 9{\cdot}81) \times 0{\cdot}75$$

$$= 56{\cdot}7 \times 10^4 + 0{\cdot}736 \times 10^4 = 57{\cdot}44 \times 10^4\,\text{N/m}^2$$

and $\qquad W = p_2 A = 57{\cdot}44 \times 10^4 \times 150 \times 10^{-4} = 8650\,\text{N}$

$$\text{Mass lifted} = \frac{W}{g} = \frac{8650}{9{\cdot}81} = \mathbf{883\,kg}$$

Pressure gauges

Atmospheric Pressure. The earth is surrounded by an atmosphere many miles high. The pressure due to this atmosphere at the surface of the earth depends upon the head of air above the surface. Atmospheric pressure at sea level is about $101{\cdot}325\,\text{kN/m}^2$, equivalent to a head of $10{\cdot}35\,\text{m}$ of water or $760\,\text{mm}$ of mercury approximately, and decreases with altitude.

Vacuum. A perfect vacuum is a completely empty space in which, therefore, the pressure is zero.

Gauge Pressure is the intensity of pressure measured above or below atmospheric pressure.

Absolute Pressure is the intensity of pressure measured above the absolute zero, which is a perfect vacuum.

Absolute pressure = gauge pressure + atmospheric pressure.

1.6 Barometers

(a) Describe, with sketches, two methods of measuring atmospheric pressure.

(b) The level of the mercury in a barometer tube is 760 mm above the level of the mercury in the bowl; what is the atmospheric pressure in N/m^2? The specific gravity of mercury is 13.6 and the specific weight of water is $9.81 \times 10^3\,\text{N/m}^3$.

Solution. (a) A *mercury barometer* in its simplest form consists of a glass tube, about $1\,\text{m}$ long and closed at one end, which is completely filled with mercury and inverted in a bowl of mercury (Fig. 1.4). A vacuum forms at the top of the tube and the atmospheric pressure acting on the surface of the mercury in the bowl supports a column of mercury in the tube of height h.

The mechanism of an *aneroid barometer* is shown in Fig. 1.5. The corrugated box is evacuated, but is prevented from collapsing by a strong spring. Variations of pressure cause the front of the box to move in or out so that the pull of the spring will just balance the force due to the atmospheric pressure. These small movements are amplified and move a pointer over a calibrated scale.

(b) If A is a point in the tube at the same level as the free surface outside, the pressure p_A at A is equal to the atmospheric pressure p at

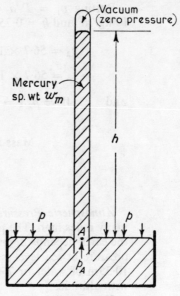

Figure 1.4

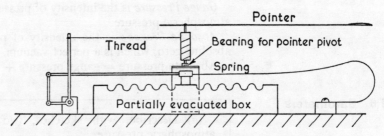

Figure 1.5

the surface because, in a fluid at rest, the pressure is the same at all points at the same level.

The column of mercury in the tube is in equilibrium under the action of the force due to p_A acting upwards and its weight acting downwards: there is no pressure on the top of the column as there is a vacuum at the top of the tube.

$$p_A \times \text{area of column } a$$

$$= \text{specific weight of mercury} \times \text{volume of column}$$

$$p_A . a = w_m . ah \quad \text{or} \quad p_A = w_m h$$

Putting $h = 760\,\text{mm} = 0.76\,\text{m}$

$$w_m = \text{specific gravity of mercury} \times \text{specific weight of water}$$

$$= 13.6 \times 9.81 \times 10^3\,\text{N/m}^3$$

$$p_A = 13.6 \times 9.81 \times 10^3 \times 0.76\,\text{N/m}^2$$

$$= \mathbf{101.3\,kN/m^2}$$

Fluid pressure measurement

1.7 Pressure tube

(*a*) Explain the use of a piezometer or pressure tube to measure the intensity of pressure in a liquid.

(*b*) A pressure tube is used to measure the pressure of oil (mass density $\rho = 640$ kg/m³) in a pipeline. If the oil rises to a height of 1·2 m above the centre of the pipe, what is the gauge pressure in N/m² at that point?

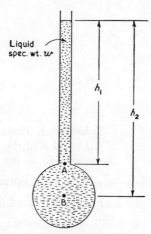

Liquid
spec. wt. w

h_1

h_2

A

B

Figure 1.6

Solution. (*a*) A piezometer or pressure tube is a vertical open-ended tube inserted into a pipe or vessel (Fig. 1.6).

Gauge pressure at A = pressure due to column of liquid of height h_1.

$$p_A = wh_1 = \rho g h_1$$

Similarly, gauge pressure at B $= p_B = \rho g h_2$

As the top of the tube is open to the atmosphere, the pressure measured is "gauge" pressure. If the liquid is moving in the pipe or vessel, the bottom of the tube must be flush with the inside of the vessel otherwise the reading will be affected by the velocity of the fluid.

(*b*) Putting $\rho g = 640$ kg/m³ and $h = 1·2$ m

$$p = 640 \times 9·81 \times 1·2 = 7·55 \text{ kN/m}^2$$

1.8 Mercury U-tube manometer

A mercury U-tube manometer is used to measure the pressure above atmospheric of water in a pipe, the water being in contact with the mercury in the left-hand limb. (*a*) Sketch the arrangement and explain its action. (*b*) If the mercury is 30 cm below A (Fig. 1.7) in the left-hand limb and 20 cm above A in the right-hand limb, what is the gauge pressure at A? Specific gravity of mercury = 13·6.

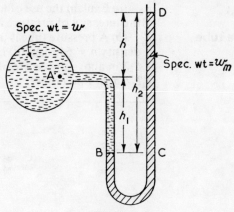

Figure 1.7

Solution. (a) The arrangement is shown in Fig. 1.7. If B is the level of the surface of the mercury in the left-hand limb and C is a point at the same level in the right-hand limb,

$$\text{pressure } p_B \text{ at B} = \text{pressure } p_C \text{ at C}$$

For the left-hand limb,

$$p_B = \text{pressure } p_A \text{ at A} + \text{pressure due to head } h_1 \text{ of water}$$
$$= p_A + wh_1$$

For the right-hand limb,

$$p_C = \text{pressure } p_D \text{ at D} + \text{pressure due to head } h_2 \text{ of mercury}$$

but p_D = atmospheric pressure = zero gauge pressure,

and so $p_C = 0 + w_m h_2 = swh_2$

where s is the specific gravity of mercury.

Since $p_B = p_C$

$$p_A + wh_1 = swh_2$$
$$p_A = swh_2 - wh_1$$

or putting $h_2 = h_1 + h$ $p_A = (s - 1) wh_1 + swh$

(b) Putting $h_1 = 30\,\text{cm} = 0.3\,\text{m}$, $h = 20\,\text{cm} = 0.2\,\text{m}$, $s = 13.6$ and $w = 9.81\,\text{kN/m}^3$.

$$p_A = (13.6 - 1) \times 9.81 \times 10^3 \times 0.3 + 13.6 \times 9.81 \times 10^3 \times 0.2\,\text{N/m}^2$$
$$= 9.81 \times 10^3(12.6 \times 0.3 + 13.6 \times 0.2) = 9.81 \times 10^3 \times 6.5$$
$$= 63.8\,\text{kN/m}^2$$

1.9 Mercury U-tube manometer

If the pressure at A in Example 1.8 is reduced by 40 kN/m² what will be the new difference in level of the mercury?

Solution. If the mercury falls x metres in the right-hand limb, it will rise x metres in the left-hand limb.

$$h_1 = (0.3 - x)\text{m and } h_2 = (0.5 - 2x)\text{m}$$

Putting $p_A = 63.8 - 40 = 23.8\,\text{kN/m}^3 = 23.8 \times 10^3\,\text{N/m}^2$, $w = 9.81 \times 10^3\,\text{N/m}^3$ and $s = 13.6$. Equating pressures in the two limbs at the new level of BC (Fig. 1.7),

$$23.8 \times 10^3 + 9.81 \times 10^3(0.3 - x) = 13.6 \times 9.81 \times 10^3(0.5 - 2x)$$

$$23.8 + 2.94 - 9.81x = 66.8 - 267x$$

$$257.2x = 40$$

$$x = 0.156\,\text{m}$$

New difference of level $= 0.5 - 2x = \mathbf{0.188\,m}$

1.10 Mercury U-tube manometer

The mercury U-tube manometer (Fig. 1.8) measures the pressure of water at A which is below atmospheric pressure. If the specific weight of mercury is 13·6 times that of water and the atmospheric pressure is 101·3 kN/m, what is the absolute pressure at A when $h_1 = 15$ cm, $h_2 = 30$ cm, and the specific weight of water is 9.81×10^3 N/m^3?

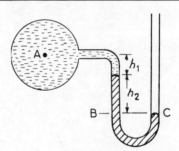

Figure 1.8

Solution. As B and C are at the same level in the same liquid at rest the pressure at B is atmospheric pressure.

$$p_B = p_C = 101.3\,\text{kN/m}^2 = 101.3 \times 10^3\,\text{N/m}^2$$

For equilibrium of the left-hand limb,

$$p_B = p_A + wh_1 + w_mh_2$$

As $w_m = 13.6w$,

$$p_A = p_B - w(h_1 + 13.6h_2)$$

Putting $h_1 = 15\,\text{cm} = 0.15\,\text{m}$, $h_2 = 30\,\text{cm} = 0.30\,\text{m}$ and $w = 9.81 \times 10^3\,\text{N/m}^3$

$$p_A = 101.3 \times 10^3 - 9.81 \times 10^3(0.15 + 13.6 \times 0.3)$$

$$= 101.3 \times 10^3 - 41.5 \times 10^3 = \mathbf{59.8 \times 10^3\,N/m^2}$$

1.11 Bourdon gauge

Solution. The mechanism of a Bourdon gauge is shown in Fig. 1.9. The fluid pressure acts on the interior of a curved tube of oval cross-section, tending to straighten it and so causing the free end to move in proportion to the pressure. This movement is amplified and used to rotate a pointer over a scale from which the pressure can be read.

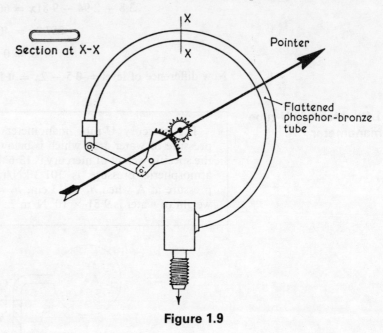

Section at X–X
Pointer
Flattened phosphor-bronze tube

Figure 1.9

Measurement of pressure differences

The pressure difference between two points in a fluid can be measured either (*a*) by measuring the pressure at each point separately and subtracting, or (*b*) by using a differential pressure gauge or manometer which measures directly the difference in pressure.

1.12 U-tube with two liquids

A U-tube manometer (Fig. 1.10) measures the pressure difference between two points A and B in a liquid of specific weight w_2. The U-tube contains mercury of specific weight w_2. Calculate the difference in pressure if $a = 1·5$ m, $b = 0·75$ m and $h = 0·5$ m, if the liquid at A and B is water ($w_1 = 9·81 \times 10^3 \, \text{N/m}^3$) and the specific gravity of mercury is 13·6 (so that $w_2 = 13·6w_1$).

Solution. Since P and Q are at the same level in the same liquid at rest, pressure p_P at P = pressure p_Q at Q.

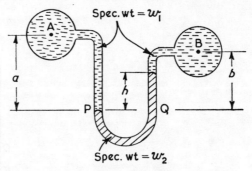

Spec. wt $= w_1$

Spec. wt $= w_2$

Figure 1.10

For the left-hand limb,
$$p_P = p_A + w_1 a$$

For the right-hand limb,
$$p_Q = p_B + w_1(b - h) + w_2 h$$

Since $p_P = p_Q$, $\quad p_A + w_1 a = p_B + w_1 b - w_1 h + w_2 h$

Pressure difference $p_A - p_B$

$= w_1(b - a) + h(w_2 - w_1)$

$= 9 \cdot 81 \times 10^3 (0 \cdot 75 - 1 \cdot 5) + 0 \cdot 5(13 \cdot 6 \times 9 \cdot 81 \times 10^3 - 9 \cdot 81 \times 10^3)$

$= 9 \cdot 81 \times 10^3 (-0 \cdot 75 + 0 \cdot 5 \times 12 \cdot 6)$

$= 9 \cdot 81 \times 10^3 \times 5 \cdot 55 = 54 \cdot 4 \times 10^3 \ \text{N/m}^2$

1.13 Inverted U-tube

An inverted U-tube manometer is used to measure the difference of water pressure between two points in a pipe. Sketch the arrangement.

If the manometer shown in Fig. 1.11 has air at the top of the tube, find the difference of pressure between point B and point A, if the mass density of water $\rho = 10^3 \ \text{kg/m}^3$, $h_1 = 60$ cm, $h = 45$ cm, and $h_2 = 180$ cm.

Solution. Fig. 1.11 shows the inverted U-tube manometer. The top of the U-tube is often filled with air, which can be forced in or released through a valve at C to control the level of the liquid in the gauge. Alternatively, the top of the U-tube may be filled with another liquid which has a lower specific gravity than the liquid at A and B and will not mix with it.

For equilibrium of the left-hand limb,
$$p_A = p_D + \rho g h_1$$

For equilibrium of the right-hand limb,
$$p_B = p_E + \rho g(h + h_2)$$
$$p_B - p_A = p_E - p_D + \rho g(h + h_2) - \rho g h_1$$

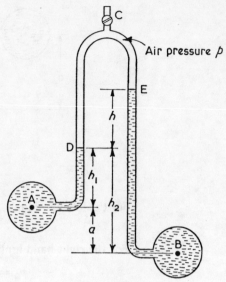

Figure 1.11

As the pressure at E and D must be equal to the pressure of the air in the top of the tube, $p_E = p_D$ and

$$p_B - p_A = \rho g(h + h_2 - h_1) = \rho g(h + a)$$

Putting $\rho = 10^3 \mathrm{kg/m^3}$, $a = h_2 - h_1 = 180 - 60 = 120\mathrm{cm} = 1.2\mathrm{m}$ and $h = 45\mathrm{cm} = 0.45\mathrm{m}$,

$$p_B - p_A = 10^3 \times 9.81(0.45 + 1.2)$$
$$= 10^3 \times 9.81 \times 1.65$$
$$= \mathbf{16.2 \times 10^3 \, N/m^2}$$

1.14 Inverted U-tube with two liquids

> The top of an inverted U-tube manometer is filled with oil of specific gravity s_o of 0·98 and the remainder of the tube with water of specific gravity s_w, of 1·01. Find the pressure difference in N/m² between two points A and B at the same level at the base of the legs when the difference in water level h is 75 mm.

Solution. Points A and B, Fig. 1.12, are some distance H below the level of the water surface in the left-hand limb. If CD is a horizontal plane through the oil-water interface in the right-hand limb, $p_C = p_D$, since pressures are equal at all points in a horizontal plane in the same liquid (oil) which is at rest.

For equilibrium of the left-hand limb,

$$p_D = p_A - s_w w H - s_o w h$$

where w is the specific weight of pure water $= 9.81 \times 10^3 \mathrm{N/m^3}$. For

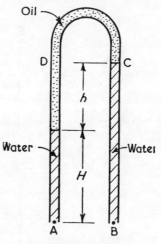

Figure 1.12

equilibrium of the right-hand limb,

$$p_C = p_B - s_w w(H + h)$$

Since $p_C = p_D$,

$$p_B - p_A = s_w w(H + h) - s_w wH - s_o wh$$
$$= wh(s_w - s_o) \text{ or } \rho gh(s_w - s_o)$$

Putting $h = 75\text{mm} = 0.075\text{m}$, $s_w = 1.01$ and $s_o = 0.98$

$$p_B - p_A = 9.81 \times 10^3 \times 0.075(1.01 - 0.98)$$
$$= 9.81 \times 10^3 \times 0.075 \times 0.03 = \textbf{22 N/m}^2$$

1.15 U-tube with enlarged ends

> The sensitivity of a U-tube gauge is increased by enlarging the ends, as shown in Fig. 1.13, and filling one side with water (specific gravity $s_w = 1$) and the other side with oil (specific gravity $s_o = 0.95$). If the area A of each enlarged end is 50 times the area a of the tube, calculate the pressure difference corresponding to a movement of 25 mm of the surface of separation between the oil and water.

Solution. As the water is denser than the oil it will fill the bottom of the U-tube.

When the pressure applied to the two limbs is the same ($p_1 = p_2$) suppose that the surface of separation between the oil and water is at level XX and that the head of oil is h.

Pressure at level XX must be the same in both limbs because the space below this plane is filled with water. For the right-hand limb, $p_x = s_o \rho gh$ and the height of the water in the left-hand limb must be

$$\frac{p_x}{s_w \rho g} = \frac{s_o}{s_w} h$$

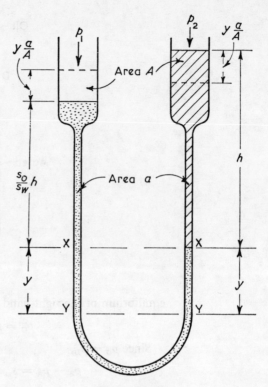

Figure 1.13

When p_2 is greater than p_1 the interface between the oil and water will move downwards in the right-hand limb a distance y to level YY.

Volume of oil withdrawn from r.-h. enlarged end $= ya$.

Fall of level in r.-h. enlarged end $= y(a/A)$.

Volume of water displaced into l.-h. enlarged end $= ya$.

Rise in level in l.-h. enlarged end $= y(a/A)$.

As the whole of the space below YY is filled with water the pressure p_Y at this level must be the same in both limbs.

For the r.-h. limb, in which the oil now extends to YY,

$$p_Y = p_2 + s_o \rho g \left(h + y - y\frac{a}{A} \right)$$

For the l.-h. limb,

$$p_Y = p_1 + s_w \rho g \left(\frac{s_o}{s_w} h + y + y\frac{a}{A} \right)$$

Therefore, $\quad p_2 - p_1 = s_w \rho g y \left(1 + \frac{a}{A} \right) - s_o \rho g y \left(1 - \frac{a}{A} \right)$

$$= \rho g y \left\{ s_w \left(1 + \frac{a}{A} \right) - s_o \left(1 - \frac{a}{A} \right) \right\}$$

Putting $y = 25\text{mm} = 0.025\text{m}$, $\rho = 10^3\text{kg/m}^3$, $s_w = 1$, $s_o = 0.95$ and $A = 50a$,

$$p_2 - p_1 = 10^3 \times 9.81 \times 0.025 \left\{ \left(1 + \frac{1}{50}\right) - 0.95 \left(1 - \frac{1}{50}\right) \right\}$$

$$= 2.45 \times 10^2 (1.02 - 0.93) = 22\,\text{N/m}^2$$

This is equivalent to a head of $\dfrac{22}{10^3 \times 9.81}\ m = 24$ mm of water.

Problems

1 Determine in (a) Newtons per square metre, (b) millibars, the increase of pressure intensity per metre of depth in fresh water. The mass density of fresh water is 1 000 kg/m^3.
Answer $(9.81 \times 10^3$ N/m^2, 98.1mb)

2 A gas holder at sea level contains gas under a pressure head equal to 9cm of water. If the mass densities of air and gas are assumed to be constant and equal to 1.28 kg/m^3 and 0.72 kg/m^3 respectively, calculate the pressure head in cm of water in a distribution main 260 m above sea level.
Answer (23.6 cm)

3 A pump delivers water against a head of 15 m of water. It also raises the water from a reservoir to the pump against a suction head equal to 250 mm of mercury. Convert these heads into N/m^2 and find the total head against which the pump works in N/m^2 and in metres of water.
Answer (147 kN/m^2, 33.4 kN/m^2, 180.4 kN/m^2, 18.4 m)

4 Two cylinders with pistons are connected by a pipe containing water. Their diameters are 75 mm and 600 mm and the face of the smaller piston is 6 m above the larger. What force on the smaller piston is required to maintain a load of 3 500 kg on the larger piston?
Answer (276 N)

5 In a hydraulic jack a force F is applied to the small piston to lift the load on the large piston. If the diameter of the small piston is 15 mm and that of the large piston is 180 mm calculate the value of F required to lift 1 000 kg.
Answer (68.2 N)

6 An hydraulic testing machine is actuated by a hand-operated pump. The hand lever has a ratio of 10 to 1 and is connected to a plunger 19 mm diam. The piston operating the testing head is 300 mm diam. Calculate the intensity of pressure in the oil and the force which must be applied to the lever when the machine exerts a force of 21 000 kgf.
Answer (2 570 $\times$ 10^3 N/m^2, 80.5 N)

7 An hydraulic press has a ram of 125 mm diam and a plunger of 12·5 mm diam. What force is required on the plunger to raise a mass of 1 000 kg on the ram? If the plunger had a stroke of 250 mm, how many strokes would be necessary to lift the weight 1m? Neglecting losses and assuming that the weight moves continuously, what power would be required to drive the plunger if the weight is lifted in 12 min?
Answer (98·1 N, 400, 13·65 W)

8 What is the atmospheric pressure in N/m² if the barometer reading is 750 mm of mercury? What would be the height of the column of water required to produce this pressure?
Answer (10^5 N/m², 10·2 m)

9 At the foot of a mountain a mercury barometer reads 740 mm and a similar barometer at the top of the mountain reads 590 mm. What is the approximate height of the mountain, assuming that the density of air is constant and equal to 1·225 kg/m³?
Answer (1 670 m)

10 The liquid in a piezometer stands 1·5 m above a point A in a pipe line. What is the pressure at A in N/m² if the liquid is (*a*) water, (*b*) oil of sp. gr. 0·85, (*c*) mercury of sp. gr. 13·6, (*d*) brine of sp. gr. 1·24? What is the pressure head in metres of each liquid?
Answer (14·7 kN/m², 12·5 kN/m², 200 kN/m², 18·3 kN/m², Head = 1·5 m)

11 In Fig. 1.7 what is the gauge pressure of the water at A if h_1 = 0·6 m and the mercury in the right-hand limb is 0·9 m above the level in the left-hand limb?
Answer (114 kN/m²)

12 In Fig. 1.14, fluid A is water and fluid B is mercury (sp. gr. 13·6). What will be the difference in level *h* if the pressure at X is 140 kN/m² and *a* = 1·5 m?
Answer (1·16 m)

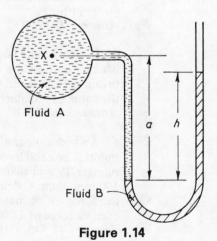

Fluid A

a   *h*

Fluid B

Figure 1.14

13 In Fig. 1.14, fluid A is oil (sp. gr. 0·8), fluid B is brine (sp. gr. 1·25). If $a = 2·5$ m and $h = 0·3$ m what are (a) the pressure head, and (b) the pressure in N/m² at X?
Answer (-2.03 m of oil, -15.9 kN/m² gauge)

14 In Fig. 1.14, fluid A is a gas (specific weight 63·5 N/m³), fluid B is water, $a = 450$ mm and $h = 150$ mm. What is the pressure in N/m² at X? What would be the error if the head a of gas is neglected?
Answer (1 467 N/m², 2·83 N/m²)

15 Assuming that the atmospheric pressure is 101·3 kN/m² find the absolute pressure at X in Fig. 1.15 when (a) fluid A is water, fluid B is mercury (sp. gr. 13·6), $a = 1$ m and $h = 0·4$ m, (b) fluid A is oil (sp. gr. 0·82), fluid B is brine (sp. gr. 1·10), $a = 20$ cm and $h = 55$ cm.
Answer ((a) 38·2 kN/m², (b) 93·8 kN/m²)

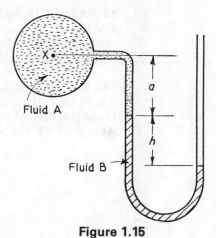

Fluid A

a

h

Fluid B

Figure 1.15

16 An oil and mercury manometer consists of a U-tube 5 mm diam with both limbs vertical. The right-hand limb is enlarged at its upper end to 25 mm diam. The enlarged end contains oil with its free surface in the enlarged portion and the surface of separation between mercury and oil is below the enlarged end. The left-hand limb contains mercury only, its upper end being open to the atmosphere. The right-hand side is connected to a vessel containing gas under pressure and the surface of separation is observed to fall 2cm. Calculate the pressure of the gas in N/m² if the specific gravity of mercury and oil are 13·6 and 0·85 respectively. The surface of the oil remains in the enlarged ends.
Answer ($5·17 \times 10^3$ N/m²)

17 In Fig. 1.16 fluid A is water and fluid B is mercury (sp. gr. 13·6). If the pressure difference between M and N is 35 kN/m², $a = 1$ m and $b = 30$ cm, what is the difference of level h?
Answer (30·7 cm)

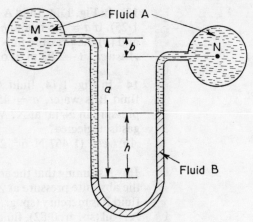

Figure 1.16

18 In Fig. 1.16 fluid A is oil (sp. gr. 0·85) and fluid B is water. If $a = 120$ cm, $b = 60$ cm and $h = 45$ cm, what is the difference of pressure in kN/m² between M and N?
Answer ($-5\cdot23$ kN/m²)

19 In Fig. 1.17 fluid A is water and fluid B is oil (sp. gr. 0·9). If h is 69 cm and z is 23 cm, what is the difference of pressure in kN/m² between M and N?
Answer ($-1\cdot57$ kN/m²)

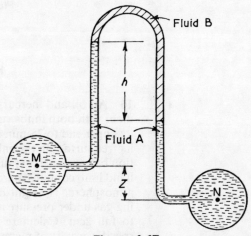

Figure 1.17

20 In Question 19 fluid A is water and fluid B is air. Assuming that the specific weight of air is negligible, what is the pressure difference in kN/m² between M and N?
Answer (4·51 kN/m²)

21 A pressure gauge consists of a U-tube with equal enlarged ends and is filled with water on one side and oil of sp. gr. 0·97 on the other, the surface of separation being in the tube below the enlarged ends. Calculate the diameter of each enlarged end if the tube diameter is 5 mm and the surface of separation moves 25 mm for a difference in pressure head of 1mm of water.
Answer (70 mm)

22 A manometer consists of an inclined glass tube which is connected to a metal cylinder standing upright: liquid fills the apparatus to a fixed zero mark on the tube when both cylinder and tube are open to the atmosphere. The upper end of the cylinder is then connected to a gas supply at a pressure p and the liquid rises in the tube. Find an expression for the pressure p in cm of water when the liquid reads y cm in the tube, in terms of the inclination θ of the tube, the specific gravity of the liquid s, and the ratio ρ of the diam of the cylinder to the diam of the tube. Hence determine the value of ρ so that the error due to disregarding the change in level in the cylinder will not exceed 0·1 per cent when $\theta = 30$ degrees.
Answer (44·6)

23 An inclined manometer is required to measure an air pressure difference of about 3 mm of water with an accuracy of ± 3 per cent. The inclined arm is 8 mm diam and the enlarged end is 24 mm diam. The density of the manometric fluid is 740 kg/m³. Find the angle which the inclined arm must make with the horizontal to achieve the required accuracy assuming that the scale can be read with a maximum error of $\pm 0·5$ mm.
Answer (7° 35′)

24 A manometer consists of a U-tube of diameter d, the upper part of each limb being enlarged to diameter D. The small tube contains liquid of relative density s_m and on top of this in each limb is a liquid of relative density s. The free surfaces are in the enlarged parts and the surfaces of separation between the two liquids of density s_m and s are in the small tube and initially level. Derive an expression for the pressure difference in a gas which, when applied across the manometer, produces a difference in the interface levels of h in the small tube, the free surfaces remaining in the enlarged ends. Also show that the sensitivity becomes greatest when the value of s approaches s_m.

In such a manometer the ratio $D/d = 5$, one of the manometer liquids is water and the other of relative density 0·95. Find the applied pressure difference required to produce a difference of interface levels h of 1 cm.

$$Answer \quad \left(p_1 - p_2 = s\rho_u gh \left\{ \left(\frac{d}{D}\right)^2 - 1 + \frac{s_m}{s} \right\}, 8·66 \text{ N/m}^2\right)$$

2

Fluid pressure on surfaces

A fluid in contact with a solid surface will exert a force on every small area of the surface equal to the product of the pressure p on the small element and its area a. Usually the pressure p will vary from point to point.

Total pressure on solid surface = sum of the forces on all these elements = Σpa.

Since there are no shear stresses in a fluid at rest the force on each element is at right angles to the solid surface. The combined action of all these elementary forces can be represented by a single resultant fluid force, or *resultant pressure*, acting at a point on the surface called the *centre of pressure*.

If the solid surface is a plane surface all the elementary forces are parallel, so that,

$$\text{Resultant force} = \text{Total pressure}$$

If the surface is curved the elementary forces will not be parallel and will have components opposing each other so that the resultant pressure will be less than the total pressure.

Note that in terms *total pressure* and *resultant pressure* the word pressure is used to mean force not intensity of pressure.

2.1 Total pressure and resultant force

Derive an expression for the total pressure on one side of a surface of area A immersed in a liquid of mass density ρ in terms of A, ρ and the vertical depth $\bar{y}$ of the centroid of the immersed surface below the free surface of the liquid.

A cylindrical tank 60 cm in diameter with its axis vertical is filled to a depth of 150 cm with water. What is the total pressure on the curved surface? What is the resultant force on this surface?

Solution. In Fig. 2.1 let a be a small area at a depth y below the free surface.

Intensity at depth $y = \rho g y$

Force on small area $= a \times \rho g y$

Total pressure = sum of forces on all such elements

$$= \Sigma a \rho g y = \rho g \Sigma a y \text{ since } \rho \text{ and } g \text{ are constant.}$$

Now Σay = first moment of area of surface about water surface

$\qquad = A\bar{y}$ where $\bar{y}$ = depth to centroid of immersed surface.

Total pressure $= \rho g A\bar{y} = A \times \rho g\bar{y}$

$\qquad$ = area of immersed surface × pressure at centroid.

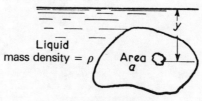

Liquid
mass density = ρ
Area
a

Figure 2.1

If cylinder is filled to a depth of 150cm, $\bar{y} = 75\text{cm} = 0.75\text{m}$

$\qquad$ Mass density of water $\rho = 10^3\text{kg/m}^3$

$\qquad$ Area of curved surface $A = \pi \times 0.6 \times 1.5 = 2.83\text{m}^2$

$\therefore$ Total pressure on curved surface $= \rho g A\bar{y}$

$\qquad\qquad\qquad = (10^3 \times 9.8) \times 2.83 \times 0.75$

$\qquad\qquad\qquad = \mathbf{20.7 \times 10^3 N}$

Since for every force on an element of area on one side of the cylinder there will be an equal and opposite force on the other side,

$\qquad\qquad$ resultant force = zero.

This is self-evident, since if there were a resultant force the cylinder would tend to move in its direction. This is clearly not the case.

Pressure on plane surfaces

2.2 Resultant fluid force on inclined plane surface

$\qquad$ A plane surface of area A is totally immersed in a liquid of specific weight w. If this surface is inclined at an angle ϕ to the horizontal and its centroid is at a vertical depth $\bar{y}$ below the free surface, derive an expression for the resultant force R on one side of the surface and also for the vertical depth D below the free surface of the centre of pressure.

Solution. Since the surface is plane, resultant force is equal to total pressure and, as shown in the previous example, since $\rho g = w$,

$\qquad\qquad$ resultant force $R = \rho g A\bar{y} = wA\bar{y}$

Since the liquid is at rest R will act at right angles to the immersed surface at C (Fig. 2.2) which is the centre of pressure. Suppose that the plane of the immersed surface AB cuts the free surface at O at the angle ϕ and that the distance from O of an element of area a of AB is L and its vertical depth below the free surface is y.

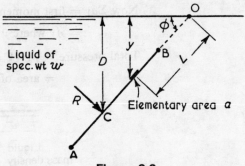

Figure 2.2

The resultant force R must be represent the combined effect of the forces on all the small area such as a which make up the immersed surface. Thus,

Moment of R about O

= Sum of moments of forces on elementary area about O.

Force on small area $a = wy \times a$

or since $y = L \sin \phi$ $= w \sin \phi \, aL$

Moment about O of force on small area $= w \sin \phi \, aL^2$

Sum of moments of all such forces about O $= w \sin \phi \, \Sigma aL^2$, since w and $\sin \phi$ are the same for all elements.

$$\text{Moment of } R \text{ about O} = R \times \text{OC} = wA\bar{y} \times \frac{D}{\sin \phi}$$

$$\therefore \qquad wA\bar{y} \times \frac{D}{\sin \phi} = w \sin \phi \, \Sigma aL^2$$

But $\Sigma aL^2 =$ 2nd moment of area of the immersed surface about O

$$= Ak_o^2$$

where $k_o =$ radius of gyration of the immersed surface about O

$$\therefore \qquad wA\bar{y} \frac{D}{\sin \phi} = w \sin \phi \, Ak_o^2$$

$$\therefore \qquad D = \sin^2 \phi \frac{k_o^2}{\bar{y}}$$

The value of k_o^2 can be found if the second moment of area of the immersed surface I_G about an axis through its centroid parallel to the water surface is known by applying the parallel axis rule.

$$I_o = I_G + A \left(\frac{\bar{y}}{\sin \phi} \right)^2$$

or $$\qquad Ak_o^2 = Ak_G^2 + A \left(\frac{\bar{y}}{\sin \phi} \right)^2$$

Thus $$\qquad D = \sin^2 \phi \frac{k_G^2}{\bar{y}} + \bar{y}$$

Note that the centre of pressure and the centre of area (or centroid) are two entirely different points. They can never coincide unless the pressure is uniform over the whole surface.

Second moments of area of common figures are given in Table 2.1.

Table 2.1 Geometrical properties of common figures

		Area A	2nd Moment of Area I_{GG} about axis GG through the centroid
Rectangle		bd	$\dfrac{bd^3}{12}$
Triangle		$\dfrac{bh}{2}$	$\dfrac{bh^3}{36}$
Circle		πR^2	$\dfrac{\pi R^4}{4}$
Semi-circle		$\dfrac{\pi R^2}{2}$	$0{\cdot}1102\ R^4$

2.3 Pressure diagram

One end of a rectangular tank is 1.5 m wide by 2 m deep. The tank is completely filled with oil of specific weight $w = 9$ kN/m³. Find by means of a pressure diagram the resultant force on this vertical end and the depth of the centre of pressure from the top.

Solution. The relation between the intensity of pressure p and the depth y is $p = wy$ which can be shown graphically as a pressure diagram (Fig. 2.3) which is a triangle in which the vertical distance represents depth and the horizontal distances represent intensity of pressure.

The area of the diagram will be the product of depth (m) and pressure intensity (N/m²) and represents to scale the resultant force R in N/m run of the immersed surface normal to the diagram.

If $H =$ depth of liquid and if the immersed surface extends from the free surface to the bottom of tank the pressure diagram is the triangle ABC.

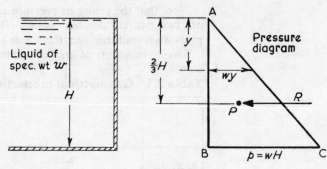

Figure 2.3

Area of pressure diagram $= \frac{1}{2} \times AB \times BC = \frac{1}{2}wH^2 =$ resultant force/unit width

Putting $w = 9\,\text{kN/m}^3$ and $H = 2\,\text{m}$

Resultant force per foot width $= \frac{1}{2}wH^2 = \frac{1}{2} \times 9 \times 2^2\,\text{kN/m}$

Width of end $= 1\cdot5\,\text{m}$

Resultant force on end $= \frac{1}{2} \times 1\cdot5 \times 9 \times 2^2 = \mathbf{27\,kN}$

The resultant R will act through the centre of area P of the pressure diagram which is $\frac{2}{3}H$ from A. Putting $H = 2\,\text{m}$,

Depth to centre of pressure $= \frac{2}{3} \times 2 = \mathbf{1\cdot33\,m}$

2.4 Inclined circular lamina

A circular lamina 125 cm in diameter is immersed in water so that the distance of its perimeter measured vertically below the water surface varies between 60 cm and 150 cm. Find the total force due to the water acting on one side of the lamina, and the vertical distance of the centre of pressure below the surface.

Solution. The derivation of the expressions for the resultant force R and the vertical distance D of the centre of pressure is given in Example 2.2. The arrangement of the lamina is shown in Fig. 2.4.

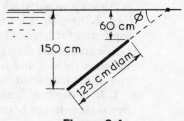

Figure 2.4

Area of lamina $A = \frac{1}{4}\pi \times (1\cdot25)^2 = 1\cdot228\,\mathrm{m}^2$

Depth to centroid $\bar{y} = \frac{1}{2}(60 + 150) = 105\,\mathrm{cm} = 1\cdot05\,\mathrm{m}$

Specific weight of water $w = 9\cdot81 \times 10^3\,\mathrm{N/m^3}$

Resultant force $R = wA\bar{y} = 9\cdot81 \times 10^3 \times 1\cdot228 \times 1\cdot05$

$$= \mathbf{12650\,N}$$

For a circular lamina $k_G{}^2 = \frac{1}{4}r^2 = \frac{1}{4}(0\cdot625)^2 = 0\cdot0976\,\mathrm{m}^2$

$$\sin\phi = \frac{150 - 60}{125} = \frac{90}{125}$$

Vertical depth of centre of pressure $D = \sin^2\phi\,\dfrac{k_G{}^2}{\bar{y}} + \bar{y}$

$$= \left(\frac{90}{125}\right)^2 \times \frac{0\cdot0976}{1\cdot05} + 1\cdot05$$

$$= 0\cdot048 + 1\cdot05 = \mathbf{1\cdot098\,m}$$

2.5 Sluice gates

A culvert draws off water from the base of a reservoir the sides of which are inclined at 80° to the horizontal (Fig. 2.5). The entrance to the culvert is closed by a circular gate 1·25 m in diameter which can be rotated about its horizontal diameter. Show that the turning moment on the gate is independent of the depth of the water if the gate is completely immersed and find the value of this moment.

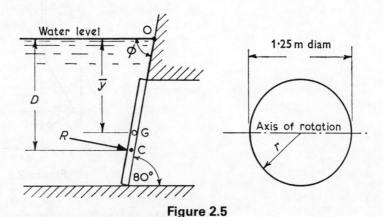

Figure 2.5

Solution. The resultant force R on the gate acts at the centre of pressure C a depth D below the surface.

The axis of rotation G passes through the centroid and is a depth $\bar{y}$ below the surface.

Turning moment on gate $= R \times CG$

$$= R \times (OC - OG) = R \left(\frac{D - \bar{y}}{\sin \phi} \right)$$

From Example 2.2 $R = \rho g A \bar{y} = \rho g \pi r^2 \bar{y}$ and $D = \sin^2 \phi \dfrac{k_G^2}{\bar{y}} + \bar{y}$

Turning moment on gate $= \rho g \pi r^2 \bar{y} \left(\dfrac{\sin^2 \phi k_G^2}{\bar{y} \sin \phi} \right)$

Putting $k_G^2 = \frac{1}{4} r^2$ for a circle, turning moment $= \frac{1}{4} \rho g \pi r^4 \sin \phi$

This is independent of $\bar{y}$ and therefore the turning moment is independent of the depth of water when completely immersed.

Substituting $\rho = 10^3 \text{kg/m}^3$, $r = 0.625\text{m}$, $\phi = 80°$,

Turning moment $= \frac{1}{4} \times 10^3 \times 9.81 \times \pi \times (0.625)^4 \times 0.985$

$$= 1160\text{N-m}$$

2.6 Sluice gates

Fig. 2.6 shows a rectangular sluice door AB hinged at the top at A and kept closed by a weight fixed to the door. The door is 120 cm wide and 90 cm long and the cg of the complete door and weight is at G, the combined weight being 1000 kgf. Find the height of water h on the inside of the door that will just cause the door to open.

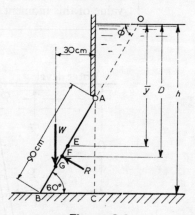

Figure 2.6

Solution. The door will just open when the moment of the resultant force R about the hinge is equal to the moment of the combined weight W about the hinge.

Vertical height of sluice $AC = 0.9 \sin 60° = 0.78\text{m}$

Depth to centre of area E of sluice $= \bar{y} = h - \frac{1}{2} \times 0.78$

$$= (h - 0.39)\text{m}$$

Resultant force $R = wA\bar{y}$

$$= (9{\cdot}81 \times 10^3) \times (1{\cdot}20 \times 0{\cdot}90) \times (h - 0{\cdot}39)$$
$$= 10{\cdot}55 \times 10^3(h - 0{\cdot}39) \text{ N}$$

Depth to centre of pressure

$$F = D = \sin^2 \phi \; \frac{k_E{}^2}{\bar{y}} + \bar{y}$$

where k_E is the radius of gyration of the gate about its centroid axis E.
For a rectangle $k_E{}^2 = d^2/12$ where $d =$ length of rectangle

$$\therefore \qquad D = \sin^2 60° \; \frac{(0{\cdot}9^2/12)}{(h - 0{\cdot}39)} + (h - 0{\cdot}39)$$

$$= \frac{0{\cdot}81}{16(h - 0{\cdot}39)} + (h - 0{\cdot}39)\, \text{metres}$$

Moment of R about hinge $A = R \times AF$

$$= R \times \frac{D - (h - 0{\cdot}78)}{\sin 60°}$$

$$= \frac{10{\cdot}55 \times 10^3(h - 0{\cdot}39)}{\sin 60°} \left\{ \frac{0{\cdot}81}{16(h - 0{\cdot}39)} + 0{\cdot}39 \right\} \text{ N-m}$$

$$= \frac{10{\cdot}55 \times 0{\cdot}81}{16 \sin 60°} + \frac{4{\cdot}12(h - 0{\cdot}39)}{\sin 60°} \text{ kN-m}$$

$$= (4{\cdot}77h - 12{\cdot}4)\, \text{kN-m}$$

Moment of W about $A = 9{\cdot}81 \times 1000 \times 0{\cdot}3\,\text{N-m} = 2{\cdot}94\,\text{kN-m}$.
When sluice is just opening

$$4{\cdot}77h - 1{\cdot}24 = 2{\cdot}94$$
$$h = \mathbf{0{\cdot}88\,m}$$

2.7 Vertical dam

The water face of a dam is in the form of a trapezium. The bottom width is a metres and the top width at the water level is $(a + b)$ metres. The face of the dam is vertical and the depth of water is h metres. Show that (a) the resultant force on the dam is $\frac{1}{6}wh^2(3a + b)$ newtons and (b) that the depth of the centre of pressure is $(4a + b)/(6a + 2b) \times h$ metres. The mass density of water is ρ kg/m³.

Solution. To determine the position of the centroid of the trapezium take moments of area about AB (Fig. 2.7).

Moment of ABCD = Moment of rectangle ABFE

$\qquad\qquad\qquad\qquad$ − Moment of triangles ADE and BFC

Area of trapezium ABCD $= \frac{1}{2}(2a + b)h = A$

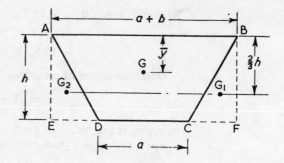

Figure 2.7

Distance of centroid from AB $= \bar{y}$

Area of rectangle ABFE $= (a + b)h$

Distance of centroid from AB $= \frac{1}{2}h$

Area of two triangles ADE and BFC $= \frac{1}{2}bh$

Distance of centroids G_1 and G_2 from AB $= \frac{2}{3}h$

Taking moments about AB

$$\tfrac{1}{2}(2a + b)h\bar{y} = (a + b)h \times \tfrac{1}{2}h - \tfrac{1}{2}bh \times \tfrac{2}{3}h$$

$$\bar{y} = h\frac{(3a + b)}{3(2a + b)}$$

(a) Resultant force $R = \rho g A \bar{y}$ (see Ex. 2.2)

$$= \rho g \times \frac{(2a + b)}{2} h \times h\frac{(3a + b)}{3(2a + b)}$$

$$= \tfrac{1}{6}\rho g h^2 (3a + b) \text{ N}$$

(b) Depth to centre of pressure $= k_{AB}^2/\bar{y}$ since surface is vertical.

Second moment of area of trapezium about AB

$\quad =$ second moment of rectangle ABFE about AB

$\qquad -$ second moment of triangles ADE and BCF

$$= \tfrac{1}{3}(a + b)h^3 - \tfrac{1}{4}bh^3 = \frac{h^3}{12}(4a + b)$$

Dividing by the area of the trapezium

$$k_{AB}^2 = \frac{h^3}{12}\frac{(4a + b)2}{(2a + b)h} = \frac{h^2}{6}\frac{(4a + b)}{(2a + b)}$$

Depth to centre of pressure $= k_{AB}^2/\bar{y}$

$$= \frac{h^2}{6}\frac{(4a + b)}{(2a + b)} \cdot \frac{3(2a + b)}{h(3a + b)} = \frac{(4a + b)}{(6a + 2b)}h \text{ metres}$$

2.8 Dock gates

A vertical dock gate is 5 m wide and has water at a depth of 7·5 m on one side and to a depth of 3 m on the other side. Find the resultant horizontal force on the dock gate and the position of its line of action. To what position does this line tend as the depth of water on the shallow side rises to 7·5 m?

Solution. In Fig. 2.8, R_1 is the resultant force on the left-hand side and R_2 is the resultant force on the right-hand side.

$$\text{Area of left-hand water face} = A_1 = BH$$

$$\text{Depth to centroid of l.-h. face} = \bar{y}_1 = \tfrac{1}{2}H$$

$$R_1 = \rho g A_1 \bar{y}_1 = \tfrac{1}{2}\rho g B H^2$$

and acts at $\tfrac{1}{3}H$ from the bottom.

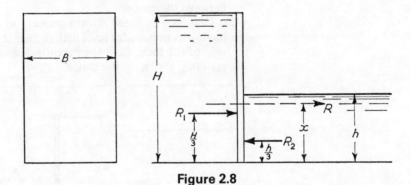

Figure 2.8

Similarly $R_2 = \tfrac{1}{2}\rho g B h^2$ and acts at $\tfrac{1}{3}h$ from the bottom. R_1 and R_2 have a resultant force R acting at a height x from the bottom. Taking moments about the bottom of the gate

$$Rx = R_1 \times \tfrac{1}{3}H - R_2 \times \tfrac{1}{3}h = \tfrac{1}{6}\rho g B H^3 - \tfrac{1}{6}\rho g B h^3$$

But
$$R = R_1 - R_2 = \tfrac{1}{2}\rho g B (H^2 - h^2)$$

$\therefore$
$$x = \frac{\tfrac{1}{6}\rho g B (H^3 - h^3)}{\tfrac{1}{2}\rho g B (H^2 - h^2)} = \frac{H^3 - h^3}{H^2 - h^2} \times \frac{1}{3}$$

$$= \frac{H^2 + Hh + h^2}{3(H + h)}$$

Putting $H = 7·5\,\text{m}$, $h = 3\,\text{m}$, $B = 5\,\text{m}$ and $\rho = 10^3\text{kg/m}^3$

$$\text{Resultant force} = R = \tfrac{1}{2}\rho g B (H^2 - h^2)$$

$$= \tfrac{1}{2}(10^3 \times 9·81 \times 5)(7·5^2 - 3^2)$$

$$= \mathbf{1160\,kN}$$

Resultant acts at x from the bottom given by

$$x = \frac{H^2 + Hh + h^2}{3(H + h)} = \frac{7\cdot5^2 + 7\cdot5 \times 3 + 3^2}{3(7\cdot5 + 3)}$$

$$= 2\cdot79\,\text{m from bottom of gate}$$

When $H = h$, $\qquad x = \dfrac{3H^2}{6H} = \tfrac{1}{2}H$

Resultant line of action tends to the mid-point of gate as level on low water side rises to 7·5 m.

2.9 Lock gates

The gates of a lock which is 7·5 m wide make an angle of 120° with each other in plan. Each gate is supported on two hinges which are situated 0·75 m and 6·25 m above the bottom of the lock. The depths of water on the two sides of the gate are 9 m and 3 m respectively. Find the force on each hinge and the thrust between the gates.

One gate is fitted with a sluice 1 m wide and 0·75 m deep having its upper edge level with the water surface on the low level side of the lock. Find the magnitude and point of action of the resultant thrust on the sluice.

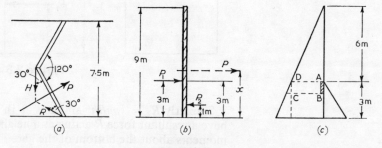

Figure 2.9

Solution. P_1 and P_2 (Fig. 2.9(b)) are the resultant water forces on the upstream and downstream side of one gate. From Ex. 2.8,

$$P_1 = \tfrac{1}{2}\rho gBH^2 = \tfrac{1}{2}(10^3 \times 9.81) \times \left(\frac{7\cdot5}{\sqrt{3}}\right) \times 9^2 = 1725\,\text{kN}$$

and will act at $\tfrac{1}{3}H = 3$ m from the bottom of the gate

$$P_2 = \tfrac{1}{2}\rho gBh^2 = \tfrac{1}{2}(10^3 \times 9.81) \times \left(\frac{7\cdot5}{\sqrt{3}}\right) \times 3^2 = 192\,\text{kN}$$

and will act at $\tfrac{1}{3}h = 1$ m from the bottom of the gate.

Resultant force on the gate $P = P_1 - P_2 = 1533\,\text{kN}$. If P acts at x from the bottom, taking moments about the bottom of the gate

$$Px = 1725 \times 3 - 192 \times 1$$
$$1533x = 4983$$
$$x = 3 \cdot 25\,\text{m}$$

The gate is in equilibrium under the water force P, the resultant R of the reactions at the top and bottom hinge (Fig. 2.9(a)), and the reaction H between the gates which acts normal to the meeting surfaces and therefore at right angles to the axis of the lock. These three forces are assumed to be in the same horizontal plane and will meet at a point since they are in equilibrium.

Resolving parallel to the gate, $H \cos 30° = R \cos 30°$, $H = R$. Resolving normal to the gate, $2R \sin 30° = P$.

$$\therefore \qquad H = R = P = 1533\,\text{kN}$$

and these forces act at $3 \cdot 25\,\text{m}$ from the bottom of the gate. The resultant reaction R is divided between the top and bottom hinges which are $6 \cdot 25\,\text{m}$ and $0 \cdot 75\,\text{m}$ from the bottom of the gate. Taking moments about the bottom hinge,

$$\text{Force on top hinge} = \frac{P(3 \cdot 25 - 0 \cdot 75)}{(6 \cdot 25 - 0 \cdot 75)} = \frac{1533 \times 2 \cdot 5}{5 \cdot 5} = \textbf{698\,kN}$$

$$\text{Force on bottom hinge} = P - 698 = 1533 - 698 = \textbf{835\,kN}$$

Fig. 2.9(c) shows the pressure diagram for the gate. Below the level AD the pressure increases equally on both sides so that below the downstream water level the pressure intensity is constant and equal to the value represented by AD. The top of the sluice is at A and the resultant force is represented by the area ABCD.

Pressure intensity represented by

$$AD = 9 \cdot 81 \times 10^3(9 - 3)\,\text{N/m}^2 = 58 \cdot 86\,\text{kN/m}^2$$
$$\text{Area of sluice} = 1 \times 0 \cdot 75 = 0 \cdot 75\,\text{m}^2$$
$$\therefore \qquad \text{Force on sluice} = 58 \cdot 86 \times 0 \cdot 75\,\text{kN}$$
$$= \textbf{44\,kN}$$

Since the pressure is uniform over the sluice, this force acts at its midpoint which is $37 \cdot 5\,\text{cm}$ below the downstream water level.

Pressure on curved surfaces

Problems involving the resultant pressure of a liquid on a curved surface can usually be solved most conveniently by considering the horizontal and vertical components of the resultant pressure separately.

2.10   Curved surface

What are the values of (i) the horizontal and (ii) the vertical components of the resultant force on one face of a curved surface immersed in a fluid, and through what points do they act? Show that if the immersed surface is part of a cylinder the resultant force will pass through the centre of curvature.

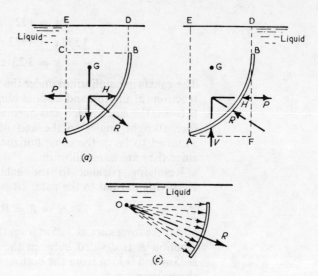

Figure 2.10

Solution. In Fig. 2.10 (*a* and *b*) AB is the immersed surface and *H* and *V* are the horizontal and vertical components of the resultant force *R* of the liquid on the surface. In Fig. 2.10(*a*) the liquid lies above the immersed surface, in Fig. 2.10(*b*) it is below this surface.

(i) Considering first the case shown in Fig. 2.10(*a*), if ACE is a vertical plane through A the section of fluid is in equilibrium horizontally under the action of *H* and the resultant force *P* of the liquid on AC, therefore *H* = *P*.

But AC is the projection of AB on a vertical plane, therefore

Horizontal component *H* = Resultant force on the projection of AB on a vertical plane.

Also for equilibrium *P* and *H* must act in the same straight line, therefore

Horizontal component *H* acts through the centre of pressure of the projection of AB on a vertical plane.

Similarly in Fig. 2.10(*b*) the section of fluid ABF is in equilibrium and as in the previous case *H* is equal to the resultant force on the projection BF of the curved surface AB on a vertical plane and acts through the centre of pressure of this projection.

(ii) In Fig. 2.10(*a*) the whole weight of the fluid ABDE lying above AB must be supported by the curved surface. For equilibrium

Vertical component *V* = weight of the fluid vertically above AB

and *V* will act vertically downwards through the centre of gravity G of the fluid.

In Fig. 2.10(*b*), if the surface AB is removed and the space ABDE is filled with the liquid, this liquid would be in equilibrium under its own weight and the vertical force *V* on the boundary AB, therefore

Vertical component *V* = weight of the fluid which *would* lie vertically above AB if AB were removed and V will act *upwards* through the centre gravity G of the imaginary fluid.

The pressure force on any small element of area of the curved surface must act at right angles to this surface and will therefore be radial if the surface is part of a cylinder. All such forces must pass through the centre of curvature O, Fig. 2.10(c), and have no moment about O. Since the resultant force R represents the combined effect of all such forces, R must also pass through the centre of curvature O.

2.11 Cylindrical gate

> A sluice gate consists of a quadrant of a circle of radius 1.5 m pivoted at its centre O (Fig. 2.11). Its centre of gravity is at G as shown. When the water is level with the pivot O, calculate the magnitude and direction of the resultant force on the gate due to the water and the turning moment required to open the gate. The width of the gate is 3 m and it has a mass of 6000 kg.

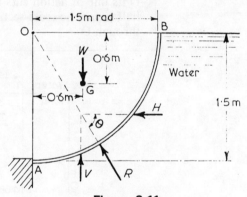

Figure 2.11

Solution. Horizontal component H of resultant pressure

= resultant pressure on projection OA of curved surface

= ρg × area of OA × depth to centroid of OA

= $(10^3 \times 9 \cdot 81) \times (3 \times 1 \cdot 5) \times 0 \cdot 75 = 33 \cdot 1 \times 10^3 \text{N}$

Vertical component V = weight of fluid which would occupy OAB

$$= \rho g \times \text{volume of cylindrical sector OAB}$$

$$= (10^3 \times 9 \cdot 81) \times (3 \times \tfrac{1}{4}\pi \times 1 \cdot 5^2)$$

$$= 52 \cdot 0 \times 10^3 \text{N}$$

Resultant force $R = \sqrt{(H^2 + V^2)}$

$$= \sqrt{(109 \cdot 5 \times 10^6 + 270 \cdot 5 \times 10^6)}$$

$$= \mathbf{61\,600\,N}$$

If θ is the angle of inclination of R to the horizontal

$$\tan \theta = 52 \cdot 0/33 \cdot 1$$

$$\theta = \mathbf{57° \ 28'}$$

Since the surface is part of a cylinder R will act through O. Therefore R has no moment about the pivot O; the only turning moment is that due to the weight of the gate.

Turning moment required to open gate $= W \times 0 \cdot 6$

$$= (6000 \times 9 \cdot 81) \times 0 \cdot 6$$

$$= \mathbf{35\,300\,N\text{-}m}$$

2.12 Parabolic surface

Fig. 2.12 shows the cross-section of a dam with a parabolic face, the vertex of the parabola being at O. The axis of the parabola is vertical and 12·5 m from the face at the water level. Estimate the resultant force in newtons per horizontal metre run due to the water, its inclination to the vertical, and how far from O its line of action cuts the horizontal OP. The centroid of the half parabolic cross-section of water is 4·68 m from the vertical through O.

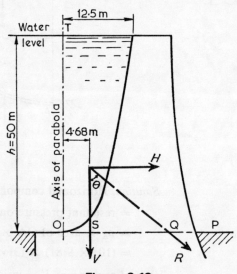

Figure 2.12

Solution. Considering one-metre-horizontal width of the dam,

Horizontal component H of water force

$$= \text{force on projection OT}$$

$$= \tfrac{1}{2}\rho gh^2 = \tfrac{1}{2}(10^3 \times 9 \cdot 81) \times 50^2 = 12 \cdot 25 \times 10^6 \text{N}$$

acting at the centre of pressure of OT, a height $\tfrac{1}{3}h = 16 \cdot 67\,$m from O.

Vertical component V = weight of water between OT and face of dam

$$= \rho g \times 1 \times \text{area of half parabola}$$
$$= (10^3 \times 9 \cdot 81) \times (\tfrac{2}{3} \times 50 \times 12 \cdot 5)$$
$$= 4 \cdot 08 \times 10^6 \, \text{N}$$

acting as given at $4 \cdot 68$ m from OT.

$$\text{Resultant force } R = \sqrt{(H^2 + V^2)}$$
$$= \sqrt{(12 \cdot 25^2 + 4 \cdot 08^2)} \times 10^6$$
$$= \mathbf{12 \cdot 9 \times 10^6 \, N}$$

Angle of inclination to vertical $\theta = \tan^{-1} \dfrac{H}{V}$

$$= \tan^{-1} \frac{12 \cdot 25 \times 10^6}{4 \cdot 08 \times 10^6}$$
$$= \mathbf{71° \ 34'}$$

The resultant R cuts OP at Q and, if V cuts OP at S,

$$OQ = OS + SQ$$
$$= OS + \tfrac{1}{3} h \tan \theta$$
$$= 4 \cdot 68 + \frac{50}{3} \times \frac{12 \cdot 25 \times 10^6}{4 \cdot 08 \times 10^6}$$
$$= 4 \cdot 68 + 50 = \mathbf{54 \cdot 68 \, m}$$

Problems

1   Find the resultant force and the centre of pressure on (a) a vertical square plate of $1 \cdot 8$ m side, and (b) a vertical plate $1 \cdot 8$ m diam. In each case the centre of the plate is $1 \cdot 2$ m below the surface of the water.
Answer $38 \cdot 1$ kN, $1 \cdot 425$ m, $29 \cdot 9$ kN, $1 \cdot 368$ m

2   A plane rectangular surface is submerged vertically in a liquid with its upper edge below the surface and parallel to it. Develop from first principles an expression for the depth to the centre of pressure.
   A vertical bulkhead has a door 2 m high and 1 m wide fastened by two hinges situated 15 cm below the top and 15 cm above the bottom of one vertical edge, and by a bolt in the centre of the other vertical edge. Calculate the forces on each hinge and the bolt when one face of the bulkhead is subjected to water pressure, the water surface being 1 m above the top of the door.
Answer $5 \cdot 52$ kN, $14 \cdot 10$ kN, $19 \cdot 62$ kN

3   A tank is 180 cm long, 90 cm deep and 90 cm wide at the top, tapering to 30 cm wide at the base. If the tank is completely filled with water, calculate (a) the total weight of water in the tank, (b) the total force exerted by the water on the bottom, (c) the total force exerted on one end.
Answer $9 \cdot 5$ kN, $4 \cdot 75$ kN, $1 \cdot 99$ kN

4 A horizontal culvert with a trapezoidal section 150 cm wide at the top, 90 cm wide at the bottom and 1·8 m high, with sides inclined equally to the vertical, connects at one end to a reservoir in which the water surface is level with the top of the culvert and is closed at the other end by a vertical bulkhead fixed by lugs at the four corners. Calculate the total force due to the water on the bulkhead and the force exerted on each fixing lug.
Answer 17·48 kN, 5·58 kN, 3·16 kN

5 A closed cylindrical tank 60 cm diam and 1·8 m deep with vertical axis, contains water to a depth of 1·2 m. Air to a pressure of 35 kN/m² above the atmosphere is pumped into the cylinder. Determine the total normal force on the vertical wall of the tank and the distance of the centre of pressure from the base.
Answer 132·3 kN, 0·856 m

6 A rectangular opening in the vertical water face of a dam is closed by a gate mounted on horizontal trunnions parallel to the longer edge of the gate and passing through the centre of the shorter vertical edge. If the water level is above the top of the gate show that the torque required to keep the gate closed is independent of the water level. Determine the magnitude of the torque when the gate is 1·25 m long by 1 m deep.
Answer 1·02 kNm

7 A circular opening A (Fig. 2.13) in the sloping wall of a reservoir is closed by a disc valve B of 70 cm diam. The disc is hinged at H and the balance weight *W* is just sufficient to hold the valve closed when the reservoir is empty. What additional mass should be placed on the balance arm 90 cm from the hinge in order that the valve shall remain closed until the water level is 60 cm above the centre of the valve?
Answer 134·5 kg

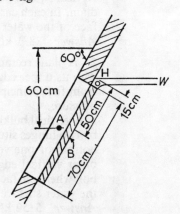

Figure 2.13

8 A rectangular sluice gate, 1·5 m wide by 1·8 m long, with its upper edge at a depth of 1·2 m below the water surface, opens by sliding on guides, the gate and guides being inclined at 45 deg. to the vertical. If the coefficient of friction between the gate and its guides is 0·12, what force parallel to the guides is required to open the gate?

Answer 5·82 kN

9 A horizontal culvert, 90 cm deep by 120 cm wide and rectangular, discharges into the sea through a wall, the face of which is inclined at an angle of 50 deg. to the horizontal. The culvert is closed by a flap valve of mass 500 kg and hinged at the top edge which just covers the opening in the wall. The cg of the valve is at its centre. If the sea covers the valve up to its hinge, to what height above the top of the culvert will fresh water be impounded before discharge occurs? Density of fresh water 1000 kg/m³ and of sea water 1025 kg/m³.

Answer 0·24 m

10 A channel is trapezoidal in section, the width at the base being 6 m and at the top 10·5 m. A dam is built across the channel with the water face at an angle of $\sin^{-1} 0·8$ to the horizontal. Find the total force on the dam and the position of the centre of pressure when the vertical depth of water is 3·6 m, the channel being full.

Answer 597 kN, 2·27 m

11 A plane surface having an area S and inclined at an angle θ to the vertical is subjected to water pressure on one side only, the centroid being at a depth of H below the water surface. Working from first principles, obtain an expression giving the depth of the centre of pressure.

The outlet end of a horizontal rectangular duct 1·2 m wide by 0·9 m deep is covered by a plane flap inclined at 40 deg to the vertical and hinged at the horizontal upper edge of the duct. The flap has a mass of 270 kg and its cg is 0·33 m horizontally from the hinge. To what height above the base must the water level in the duct rise so as just to cause the flap to open?

Answer 0·332 m

12 The face of a dam is vertical to a depth of 7·5 m below the water surface and then slopes at 30 deg to the vertical. If the depth of water is 16·5 m specify completely the resultant force per metre run acting on the whole face.

Answer (1·475 × 10⁶N, at 24° 36′ to horizontal through a point 5·5 m above base and 2·93 m from vertical face)

13 Develop from first principles a general expression for the position of the centre of pressure of a fluid on one side of a sloping flat plate completely immersed in the fluid.

A rectangular opening in the sloping side of a reservoir containing water is 90 cm by 60 cm with the 60 cm side horizontal. It is closed by means of a gate as shown in Fig. 2.14. The gate is hinged at the top edge and kept closed against the water pressure partly by its own weight and partly by a weight W on a lever arm.

Assuming the gate to be a uniform flat plate of mass 45 kg, and neglecting the weight of the lever arm, calculate the mass of the counterweight W required so that the gate will commence to open

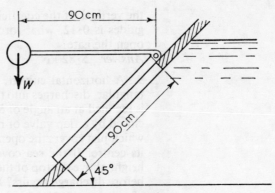

Figure 2.14

when the water level reaches a height of 30 cm above the top of the gate.
Answer   161 kg

14   A masonry dam 6 m high has the water level with the top. Assuming that the dam is rectangular in section and 3 m wide, determine whether the dam is stable against overturning and whether tension will develop in the masonry joints. Density of masonry is 1760 kg/m³.
Answer   Stable, tension on water face

15   The angle between a pair of lock gates is 140 deg and each gate is 6 m high and 1·8 m wide, supported on hinges 0·6 m from the top and bottom of the gates. If the depths of water on the upstream and downstream sides are 5 m and 1·5 m respectively, estimate the reactions at the top and bottom hinges.
Answer   73 kN, 221 kN

16   The end gates of a lock are 4·8 m high and when closed include an angle of 120 deg. The width of the lock is 6 m. Each gate is carried on two hinges placed at the top and bottom of the gate. If the water levels are 4·5 m and 3 m on the up and down stream sides respectively, determine the magnitude of the force on the hinges due to water pressure.
Answer   75·6 kN, 115·1 kN

17   A dam has its water face in the shape of a circular arc as shown in Fig. 2.15. Calculate the resultant force on the curved surface per metre run and its inclination to the horizontal.
Answer   860 kN, 35 deg

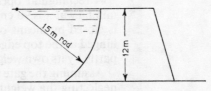

Figure 2.15

18 A sector-shaped sluice gate having a radius of curvature of 5·4 m is as shown in Fig. 2.16. The centre of curvature C is 0·9 m vertically below the lower edge A of the gate and 0·6 m vertically above the horizontal axis passing through O about which the gate is constructed to turn. The mass of the gate is 3000 kg per m run and its centre of gravity is 3·6 m horizontally from the centre O. If the water level is 2·4 m above the lower edge of the gate find, per metre run, (*a*) the resultant force acting on the axis at O, (*b*) the resultant moment about O.
Answer   36·2 kN, 89 kN-m

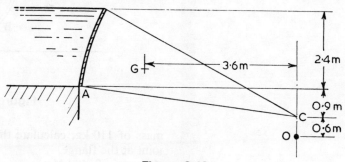

Figure 2.16

19 A sector gate is constructed with a radius of 6 m and subtends an angle of 25 deg. The sector takes the pressure of water which is level with the upper edge. The line from the hinge to the lower edge is inclined upwards from the horizontal at 10 deg. Calculate the resultant force at the hinge per metre width of gate, also the horizontal component of this force.
Answer   29·68 kN, 28·25 kN

20 A special sluice gate is of the form shown in Fig. 2.17. A is a flat gate freely suspended by a hinge at C and hanging vertically under its own weight in which position it makes contact with B, a sector gate of 0·75 m radius carried by a shaft through the centre of curvature at D. The sector gates weighs 500 kgf per metre run, its cg being at G as shown. If the water level is 1·65 m above the floor, calculate per metre width the force on the hinge pin D in magnitude and direction, and the torque on the shaft D required to open the gate.
Answer   11·2 kN, 28° 28′, 2·21 kN-m

21 Explain a method by which the resultant force and the centre of pressure on any non-planar surface immersed in a fluid can be calculated.

A flat circular base plate rests firmly on a horizontal foundation. Placed above it is a hemispherical container of radius 60 cm with a flange bolted to the flat plate. The centre of the hemisphere is at the surface of the plate. The container is filled with water to a depth of 45 cm and in the air space there is a pressure of 35 kN/m² above atmosphere. If the container has a

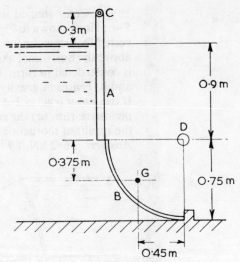

Figure 2.17

mass of 110 kg, calculate the total force tending to break the joint at the flange.

Answer   39·48 kN

22   The half-section of a ship is shown in Fig. 2.18. The side is vertical to a depth of 60 cm below the waterline and then curves to the centre-line in the form of a parabolic arc, the axis of the parabola being a horizontal line 60 cm below the waterline. Determine the magnitude and direction of the resultant hydrostatic thrust per metre of length on the half-section, and where the line of action of this resultant cuts the vertical centre-line. Density of sea water = 1025 kg/m³.

Answer   33·8 kN, 28° 52′ to vertical, 2·3 m above bottom

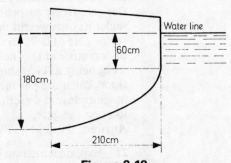

Figure 2.18

23   A stream is spanned by a bridge which is a single masonry arch in the form of a parabola, the crown being 2·4 m above the springings which are 9 m apart. Measured in the direction of the stream the overall width is 6·3 m. During a flood the stream rises to a level of 1·8 m above the springings. Assuming that the arch remains watertight, calculate the force tending to lift the bridge from its foundations.

Answer   222 kN

24   The face of a dam, Fig. 2.19, is curved according to the relation $y = x^2/2 \cdot 4$ where y and x are in metres. The height of the free surface above the horizontal plane through A is $15 \cdot 25$ m. Calculate the resultant force F due to the fresh water acting on unit breadth of the dam, and determine the positon of the point B at which the line of action of the force cuts the horizontal plane through A.

Answer   1290 kN/m, $11 \cdot 84$ m

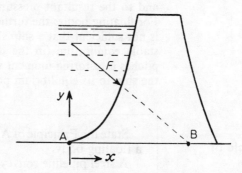

3

Buoyancy and stability of floating bodies

Most problems concerned with totally or partially immersed bodies are problems of equilibrium between forces due to the weight of the body and to the resultant pressure of the fluid on the surface of the body. For floating bodies the further problem of stability will usually arise. It is not sufficient that a ship should be in equilibrium, it must also be in stable equilibrium (in the desired position) so that when it rolls or pitches a restoring moment will be produced which will tend to return the ship to its equilibrium position.

Buoyancy

3.1 Principle of Archimedes

State the Principle of Archimedes and explain its application to a floating body.

A steel pipeline conveying gas has an internal diam of 120 cm and an external diam of 125 cm. It is laid across the bed of a river, completely immersed in water and is anchored at intervals of 3 m along its length. Calculate the buoyancy force in newtons per metre run and the upward force in newtons on each anchorage. Density of steel = 7900 kg/m³, density of water 1000 kg/m³.

Solution. The Principle of Archimedes states that the upthrust (upward vertical force due to the fluid) on a body immersed in a fluid is equal to the weight of the fluid displaced.

The upthrust will act through the centre of gravity of the displaced fluid, which is called the *centre of buoyancy*.

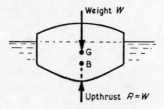

Figure 3.1

A floating body (Fig. 3.1) will be in equilibrium under the action of the weight W acting vertically at its centre of gravity G and the upthrust R acting vertically upwards through the centre of buoyancy B. For equilibrium,

$$\text{Weight of body } W = \text{Upthrust } R$$
$$= \text{Weight of liquid displaced}$$

and R and W must act in the same vertical straight line.

$$\text{Buoyancy force/metre run} = \text{upthrust/metre run}$$
$$= \text{weight of water displaced/metre run}$$
$$= (1000 \times 9{\cdot}81) \times \tfrac{1}{4}\pi(1{\cdot}25)^2$$
$$= \mathbf{12\,150\,N/m}$$

Since the anchorages are 3 m apart

Upward force on anchorage

$$= (\text{Buoyancy force} - \text{weight}) \text{ for 3 m of pipe}$$
$$\text{Weight of 3 m of pipe} = 3 \times (7900 \times 9{\cdot}81) \times \tfrac{1}{4}\pi(1{\cdot}25^2 - 1{\cdot}20^2)$$
$$= 183\,000 \times 2{\cdot}45 \times 0{\cdot}05 = 22\,500\,N$$
$$\text{Buoyancy force on 3 m of pipe} = 3 \times 12\,150 = 36\,450\,N$$
$$\text{Upward force on anchorage} = 36\,450 - 22\,500$$
$$= \mathbf{13\,950\,N}$$

3.2 Upthrust on immersed body

Show from consideration of the force exerted on an immersed surface in a fluid that the resultant force on a body immersed in a fluid is vertical and equal to the weight of fluid displaced.

A ship floating in sea water displaces 115 m³. Find (a) the weight of the ship if sea water has a density of 1025 kg/m³, and (b) the volume of fresh water of density 1000 kg/m³ which the ship would displace.

Solution. In Fig. 3.2 ABCDE is an irregular solid immersed in a fluid. From Example 2.10 it has been shown that

1. Horizontal force on a curved surface immersed in a fluid is equal to the force on the projection of the surface on a vertical plane.

2. Vertical force on a curved surface immersed in a fluid is equal to the weight of the fluid which lies or would lie vertically above it.

Considering any vertical plane drawn through the immersed body, the projected area of the two halves of its surface on this plane will be equal and the horizontal forces on the two sections of the surface will therefore be equal and opposite. There is therefore no horizontal force on the body, and the resultant force due to buoyancy must be vertical.

In Fig. 3.2, if ADC is a horizontal plane:

$$\text{Upthrust on body} = \text{upward force on lower surface ADEC}$$
$$- \text{downward force on upper surface ABCD}$$
$$= \text{wt of volume of liquid AECDHGF}$$
$$- \text{wt of volume of liquid ABCDGHF}$$

= wt of volume of liquid ABCDE

= wt of fluid displaced by the body

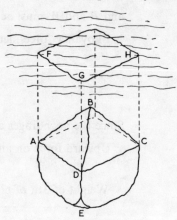

Figure 3.2

(a)   Weight of ship = weight of sea water displaced

∴   mass of ship = mass of sea water displaced

= 1025 × 115 = **118000 kg**

(b)   Mass of fresh water displaced = mass of ship

= 118000 kg

Volume of fresh water displaced = 118000/1000 = **118 m³**

3.3 Stability and metacentre

(a) What are the three conditions in which a solid body can be in equilibrium?

(b) Define the term "metacentre" and show how the stability of a floating body depends upon the position of the metacentre and the centre of gravity.

(c) A vessel has a displacement of 2 500 000 kg of fresh water. A mass of 20 000 kg moved 9 m across the deck causes the lower end of a pendulum 3 m long to move 23 cm horizontally. Calculate the transverse metacentric height.

Solution. (a) The three possible conditions of equilibrium are

1. *Stable equilibrium.* A small displacement from the equilibrium position produces a righting moment tending to restore the body to the equilibrium position.

2. *Unstable equilibrium.* A small displacement produces an overturning moment tending to displace the body further from its equilibrium position.

3. *Neutral equilibrium.* The body remains at rest in any position to which it may be displaced.

Fig. 3.3 shows these three conditions for a cone placed on a horizontal surface.

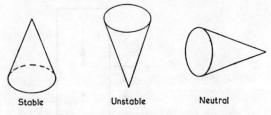

Stable   Unstable   Neutral

Figure 3.3   Conditions of equilibrium

(b) Fig. 3.4(a) shows a solid body floating in equilibrium. The weight W acts through the centre of gravity G and the upthrust R acts through the centre of buoyancy B, which is the centre of gravity of the displaced fluid, both R and W acting in the same straight line. When the body is

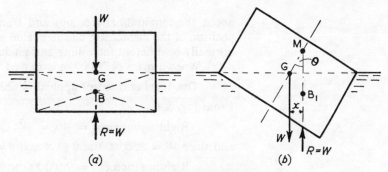

Figure 3.4   Stable equilibrium

displaced from its equilibrium, W continues to act at G. The volume of liquid displaced remains constant (since $R = W$) but usually the shape of this volume will change and the position of its centre of gravity and centre of buoyancy B will move relative to the body. Thus in Fig. 3.4(a) the displaced fluid is rectangular in section while in Fig. 3.4(b) it is triangular and the centre of buoyancy moves to B_1. As a result R and W are no longer in the same straight line but are equal and opposite parallel forces producing a turning moment Wx which is a righting moment in Fig. 3.4 and an overturning moment in Fig. 3.5.

The *metacentre* M is the point at which the line of action of R for the displaced position cuts the original vertical through the centre of gravity of the body G.

The *metacentric height* is the distance GM.

For a small angle of tilt θ, Righting Moment = $Wx = W \cdot \text{GM} \cdot \theta$ (since for small angles $\tan \theta = \sin \theta = \theta$ radians).

Comparing Figs. 3.4 and 3.5, it can be seen that:

1. If M lies above G a righting moment is produced, GM is regarded as positive, and equilibrium is stable;

2. If M lies below G an overturning moment is produced, GM is regarded as negative, and equilibrium is unstable;

3. If M and G coincide the body is in neutral equilibrium.

Since a floating body can be displaced from its equilibrium position in any direction, it is usual in the case of a ship to consider displacements

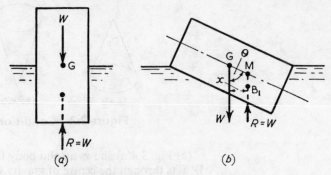

Figure 3.5 Unstable equilibrium

about the longitudinal (rolling) and transverse (pitching) axes; the position of the metacentre and the value of the metacentric height will normally be different for rolling and pitching.

(c) When a mass of 20000 kg is moved 9 m across deck

$$\text{Overturning moment applied} = (20000 \times 9{\cdot}81) \times 9\,\text{N/m}$$

From Fig. 3.4

$$\text{Righting moment} = Wx = W\,.\,\text{GM}\,.\,\theta$$

and since $W = 2\,500\,000 \times 9{\cdot}81\,\text{N}$ and $\theta = 0{\cdot}23/3$

$$\text{Righting moment} = 2\,500\,000 \times 9{\cdot}81 \times 0{\cdot}23/3 \times \text{GM}$$

Since the vessel is in equilibrium in the tilted position,

$$\text{Overturning moment} = \text{righting moment}$$

$$20000 \times 9{\cdot}81 \times 9 = 2\,500\,000 \times 9{\cdot}81 \times 0{\cdot}23/3 \times \text{GM}$$

$$\text{Metacentric height} = \text{GM} = \frac{3 \times 20 \times 9}{2\,500 \times 0{\cdot}23}$$

$$= \mathbf{0{\cdot}94\,m}$$

3.4   BM = I/V

Show that, if B is the centre of buoyancy and M is the metacentre for rolling of a partially immersed floating body, BM $= I/V$ where I is the second moment of area of the surface of flotation about the longitudinal axis, and V is the immersed volume.

The shifting of a portion of cargo of mass 25 000 kg, through a distance of 6 m at right angles to the vertical plane containing the longitudinal axis of a vessel, causes it to heel through an angle of 5 deg. The displacement of the vessel is 5000 metric tons and the value for I is 5840 m⁴. The density of sea water is 1025 kg/m³.

Find (a) the metacentric height and (b) the height of the centre of gravity of the vessel above the centre of buoyancy.

Solution. In Fig. 3.6 B is the centre of buoyancy when the vessel is in its equilibrium positon and AC the corresponding water plane. When

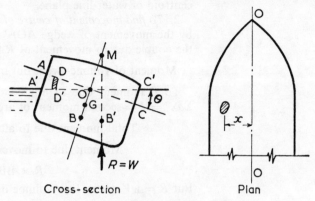

Cross-section   Plan

Figure 3.6

the vessel is tilted through an angle θ the centre of buoyancy will move to B′ as a result of the alteration of the shape of the displaced fluid, and A′C′ is the corresponding water plane.

For small angles of tilt θ, $BM = \dfrac{BB'}{\theta}$.

The centre of buoyancy is the centre of gravity of the displaced fluid. When the vessel tilts, the shape of the volume of displaced fluid is altered by the removal of a wedge AOA′ and the addition of a wedge COC′.

If the position of the axis O can be found, the distance BB′ can be calculated in the usual manner for centres of gravity and hence BM can be found.

1. *To find position of O.* When the vessel is tilted the weight of liquid displaced remains unchanged, therefore

$$\text{weight of wedge AOA}' = \text{weight of wedge COC}'$$

If a is a small area in the water-line plane at a distance x from the axis of rotation OO, it will generate a small volume when the vessel tilts.

$$\text{Volume swept out by } a = DD' \times a = ax\theta$$

Summing and multiplying by the specific weight of the liquid $w = \rho g$

$$\text{Weight of wedge AOA}' = \sum_{x=0}^{x=AO} wax\,\theta$$

Similarly,

$$\text{Weight of wedge COC}' = \sum_{x=0}^{x=CO} wax\,\theta$$

Since there is no change in displacement

$$w\theta \sum_{x=0}^{x=AO} ax = w\theta \sum_{x=0}^{x=CO} ax$$

$$w\theta \sum_{x=AO}^{x=CO} ax = 0$$

Since w and θ are not zero, $\Sigma ax = 0$. But Σax is the first moment of area of water-line plane about OO, therefore axis OO passes through centroid of water-line plane.

2. *To find movement of centre of buoyancy BB'.* The couple produced by the movement of wedge AOA' to position COC' must be equal to the couple due to movement of R from B to B'.

Moment of volume swept out by area a about OO $= w\theta ax \times x$.

Total moment due to altered displacement $= w\theta\Sigma ax^2$, or putting $\Sigma ax^2 = I =$ second moment of area of the waterline plane about OO,

$$\text{Total moment due to altered displacement} = w\theta I$$

$$\text{Moment due to movement of } R = R \times \text{BB}'$$

$$\therefore \qquad\qquad R \times \text{BB}' = w\theta I$$

But $R = wV$ where $V =$ volume displaced by the fluid. Thus

$$\text{BB}' = \frac{\theta I}{V}$$

and

$$\text{BM} = \frac{\text{BB}'}{\theta} = \frac{I}{V}$$

(a) $\qquad\qquad$ Overturning moment $=$ righting moment.

If $P =$ weight of cargo moved, $x =$ distance moved, $W =$ weight of vessel and $\theta =$ angle of tilt in radians

$$Px = W \cdot \text{GM} \cdot \theta$$

$$(25000 \times 9\cdot81) \times 6 = (5000 \times 10^3 \times 9\cdot81) \cdot \text{GM} \times \frac{5}{360} \times 2\pi$$

$$\text{Metacentric height} = \text{GM} = \frac{25 \times 6 \times 360}{5000 \times 5 \times 2\pi} = \mathbf{0\cdot344\,m}$$

(b) Volume of water displaced $= V = \dfrac{5000 \times 10^3}{1025} = 4880\,\text{m}^3$

Second moment of area of water plane $= I = 5840\,\text{m}^4$

$$\text{BM} = \frac{I}{V} = \frac{5840}{4880} = 1\cdot195\,\text{m}$$

Distance from centre of gravity to centre of buoyancy

$$= \text{BG} = \text{BM} - \text{GM}$$

$$= 1\cdot195 - 0\cdot344 = \mathbf{0\cdot851\,m}$$

3.5 Stability of floating cone

State the conditions to be met to ensure the stable equilibrium of a body partly immersed in a liquid.

A right solid cone with apex angle equal to 60 deg. is of density k relative to that of the liquid in which it floats with apex downwards. Determine what range of k is compatible with stable equilibrium.

Solution. For a body to be in stable equilibrium when partly immersed in a liquid:

1. Weight of body = weight of liquid displaced,
2. The metacentric height must be positive.

If R (Fig. 3.7) is the radius of the base of the cone, D the vertical height, r = radius of the water-line plane and d the depth immersed

$$r = \frac{1}{\sqrt{3}}d \quad \text{and} \quad R = \frac{1}{\sqrt{3}}D$$

G is the centre of gravity of the cone so that $OG = \frac{3}{4}D$.
B is the centre of buoyancy, therefore $OB = \frac{3}{4}d$.
For the body to float

$$\text{Weight of cone} = \text{weight of liquid displaced}$$
$$\tfrac{1}{3}\pi R^2 D \times k\rho g = \tfrac{1}{3}\pi r^2 d \times \rho g$$

where ρ = mass density of liquid. Substituting for r and R,

$$\frac{k\pi D^3}{9} = \frac{\pi d^3}{9}$$
$$d = k^{1/3}D$$

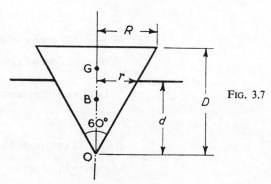

Figure 3.7

FIG. 3.7

Thus
$$OB = \tfrac{3}{4}d = \tfrac{3}{4}k^{1/3}D$$

Also
$$BM = \frac{I}{V} = \frac{\tfrac{1}{4}\pi r^4}{\tfrac{1}{9}\pi d^3} = \frac{9r^4}{4d^3} = \tfrac{1}{4}d = k^{1/3}\tfrac{1}{4}D$$

$$BG = OG - OB = \tfrac{3}{4}D(1 - k^{1/3})$$

For stability the metacentric height $GM = BM - BG$ must be

positive, so that BM must be greater than BG or

$$k^{1/3}\tfrac{1}{4}D > \tfrac{3}{4}D(1 - k^{1/3})$$
$$k^{1/3} > 3 - 3k^{1/3}$$
$$4k^{1/3} > 3$$
$$k > 0\cdot421$$

Also if the body is not to sink k must be less than 1.

$$\text{Permissible range of } k = \textbf{0\cdot421 to 1}$$

3.6 Buoy with anchor

A cylindrical buoy $1\cdot35$ m in diam and $1\cdot8$ m high has a mass of 770 kg. Show that it will not float with its axis vertical in sea water of density 1025 kg/m³. If one end of a vertical chain is fastened to the base, find the pull required just to keep the buoy vertical. The centre of gravity of the buoy is $0\cdot9$ m from its base.

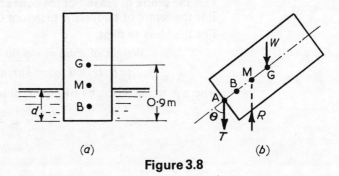

(a)   *(b)*

Figure 3.8

(a) *Without the anchor chain, Fig. 3.8(a).*

$$\text{Volume of water displaced} = V = 770/1025 = 0\cdot75\,\text{m}^3$$

$$\text{Depth of buoy immersed} = d = \frac{V}{\pi r^2} = \frac{0\cdot75}{\pi \times (0\cdot675)^2}$$

$$= 0\cdot524\,\text{m}$$

$$\text{Height of centre of buoyancy above base} = \tfrac{1}{2}d = 0\cdot262\,\text{m}$$

$$\text{Height of centre of gravity above base} = 0\cdot9\,\text{m}$$

$$\text{BG} = 0\cdot9 - 0\cdot262 = 0\cdot638\,\text{m}$$

Also

$$\text{BM} = \frac{I}{V} = \frac{\pi \times \tfrac{1}{4}r^4}{V} = \frac{\tfrac{1}{4}\pi(0\cdot675)^4}{0\cdot75} = 0\cdot218\,\text{m}$$

$$\text{Metacentric height GM} = \text{BM} - \text{BG} = -0\cdot42\,\text{m}$$

Negative metacentric height indicates that the buoy is unstable.

(b) *With the anchor chain, Fig. 3.8(b).* If $T = $ pull in the chain in newtons,

$$\text{New upthrust } R = T + W$$

$$\text{New displacement volume} = \frac{R}{\rho g} = \frac{R}{1025 \times 9.81} \text{ m}^3$$

$$\text{New draught} = \frac{R}{1025 \times 9.81 \times \pi \times (0.675)^2} = \frac{R}{14\,400} \text{ m}$$

$$\text{Height of centre of buoyancy above A} = \frac{R}{28\,800} \text{ m}$$

$$\text{BM} = \frac{I}{V} = \frac{\frac{1}{4}\pi(0.675)^4}{R/(1025 \times 9.81)} = \frac{1635}{R} \text{ m}$$

$$\text{AM} = \text{AB} + \text{BM} = \frac{R}{28\,800} + \frac{1635}{R} \text{ m}$$

$$\text{AG} = 0.9 \text{ m} \quad \text{and} \quad \text{GM} = 0.9 - \frac{R}{28\,800} - \frac{1635}{R} \text{ m}$$

For equilibrium, taking moments about G,

$$0.9T = R\left(0.9 - \frac{R}{28\,800} - \frac{1635}{R}\right)$$

$$0.9(R - 770 \times 9.81) = 0.9R - \frac{R^2}{28\,800} - 1635$$

$$R^2 = 28\,800(6800 - 1635) = 149 \times 10^6$$

$$R = 12\,200 = 770 \times 9.81 + T \text{ newtons}$$

$$\text{Tension in chain} = T = 12\,200 - 7560 = \mathbf{4640\,N}$$

3.7 Angle of tilt

An empty tank is rectangular in plan, side elevation and end elevation, and is made of sheet steel. The top is closed by a sheet of the same metal and the dimensions of the tank are − length 120 cm, breadth 66 cm and height 60 cm. If the sheet steel weighs 369 N/m² and the tank is allowed to float in fresh water with the 60 cm edges vertical, prove that the equilibrium is unstable, and find the inclination of these edges to the vertical after the tank has heeled over to a stable position. If a formula is used for the heeling condition it must be proved.

Solution. (*a*) *In the vertical position*

Weight of tank

$$= 369 \times (2 \times 1.2 \times 0.66 + 2 \times 1.2 \times 0.6 + 2 \times 0.66 \times 0.6)$$

$$= 369 \times 3.82 = 1410 \text{ N}$$

$$\text{Displacement} = V = \frac{1410}{9.81 \times 10^3} = 0.1435 \text{ m}^3$$

$$\text{Draught} = \frac{0.1435}{1.2 \times 0.66} = 0.182\,\text{m}$$

Height of centre of buoyancy above bottom $= \frac{1}{2} \times 0.182 = 0.091\,\text{m}$

Height of centre of gravity of tank $= \frac{1}{2} \times 0.6 = 0.3\,\text{m}$

Distance of cg above centre of buoyancy $= \text{BG} = 0.209\,\text{m}$

$$\text{BM} = \frac{I}{V} = \frac{1.2 \times (0.66)^3/12}{0.1435} = 0.200\,\text{m}$$

Metacentric height $\text{GM} = \text{BM} - \text{BG} = 0.200 - 0.210$
$$= -0.010\,\text{m}$$

The tank is therefore unstable with 60 cm edge vertical.

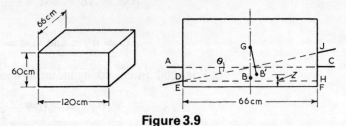

Figure 3.9

(b) If the 60 cm edge makes an angle θ to the vertical so that the water plane moves from AC to DJ (Fig. 3.9), the centre of buoyancy B will move to B' a height z from the bottom. The mean draught is unchanged and therefore

$$\text{Area DEFJ} = 0.180 \times 0.66 = 0.119\,\text{m}^2$$
$$\text{ED} = 0.182 - 0.33 \tan \theta$$
$$\text{FJ} = 0.182 + 0.33 \tan \theta$$
$$\text{JH} = \text{FJ} - \text{ED} = 0.66 \tan \theta$$

To find z take moments of area about EF

$$\text{Area DEFJ} \times z = \text{moment of DEFH} + \text{moment of DHJ}$$
$$0.119z = 0.66(0.182 - 0.33 \tan \theta)^2 \times \tfrac{1}{2} + \{0.66 \times 0.66 \tan \theta$$
$$\times \tfrac{1}{2}(0.182 - 0.33 \tan \theta + 0.22 \tan \theta)\}$$
$$= 0.33(0.0324 + 0.036 \tan{}^2\theta)$$
$$z = 0.0898 + \tan{}^2\theta$$

and $\qquad \text{B'G} = (0.3 - z) \sec \theta$

$$\text{New } I \text{ of water plane} = \frac{1.2 \times (\text{DJ})^3}{12}$$
$$= \frac{(0.66 \sec \theta)^3}{10} = 0.0288 \sec{}^3\theta$$

If M' is the new metacentre

$$\text{B'M'} = \frac{I}{V} = \frac{0.0288 \sec{}^3\theta}{0.1435} = 0.202 \sec{}^3\theta$$

The tank will just be in equilibrium when B'G = B'M'

$$(0\cdot3 - z) \sec\theta = 0\cdot202 \sec {}^2\theta$$

$$0\cdot3 - 0\cdot0898 - \tan {}^2\theta = 0\cdot202 \sec {}^2\theta$$

$$= 0\cdot202 + 0\cdot202 \tan {}^2\theta$$

$$1\cdot202 \tan {}^2\theta = 0\cdot009$$

$$\tan\theta = 0\cdot0866$$

$$\theta = 5\deg$$

3.8 Time of oscillation

A floating body oscillates about its equilibrium position. Show that the periodic time of rolling is given by $t = 2\pi\sqrt{(k^2/gm)}$ where m is the metacentric height and k the radius of gyration about the centre of gravity.

A ship has a displacement of 5000 metric tons. The second moment of area of the water-line section about a fore-and-aft axis is 12 000 m^4 and the centre of buoyancy is 2 m below the cg. The radius of gyration is 3·7 m. Calculate the period of oscillation. Sea water has a density of 1025 kg/m^3.

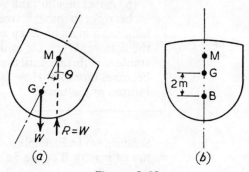

Figure 3.10

Solution. From Fig. 3.10(*a*) for a displacement θ

$$\text{Righting moment} = T = W\,.\,\text{GM}\,.\,\theta = Wm\theta$$

$$\text{Angular acceleration due to } T = \frac{T}{I} = \frac{Wm\theta}{(W/g)k^2} = \frac{m\theta g}{k^2}$$

Since the acceleration is proportional to θ the motion is simple harmonic motion.

$$\text{Periodic time } t = 2\pi\sqrt{\frac{\text{displacement}}{\text{acceleration}}} = 2\pi\sqrt{\frac{\theta}{m\theta g/k^2}}$$

$$t = 2\pi\sqrt{\frac{k^2}{gm}}$$

In Fig. 3.10(*b*)

$$BG = 2\,\text{m}$$

$$\text{Displacement volume } V = \frac{5000 \times 1000}{1025} = 4880\,\text{m}^3$$

$$BM = \frac{I}{V} = \frac{12000}{4880} = 2\cdot45\,\text{m}$$

$$GM = BM - BG = 2\cdot45 - 2 = 0\cdot45\,\text{m}$$

$$\text{Periodic time} = 2\pi\sqrt{\frac{3\cdot7^2}{0\cdot45 \times 9\cdot81}} = \mathbf{11\cdot1\,s}$$

3.9 Vessel containing liquid

(*a*) A tank containing liquid is tilted about its longitudinal axis through a small angle θ. Show that the centre of gravity G of the liquid in the tank will move to a position G′ such that

$$GG' = \theta\frac{I}{V}$$

where I is the second moment of area of the surface of the liquid about the longitudinal axis and V is the volume of liquid in the tank.

(*b*) A rectangular tank 90 cm long and 60 cm wide is mounted on bearings so that it is free to turn about a longitudinal axis. The tank has a mass of 68 kg and its centre of gravity is 15 cm above the bottom. When the tank is slowly filled with water it hangs in stable equilibrium until the depth of water is 45 cm, after which it becomes unstable. How far is the axis of the bearings above the bottom of the tank?

Solution. (*a*) In Example 3.4 it was shown that, for angle of tilt θ, the centre of gravity B of the fluid displaced by a vessel moved to B′ such that

$$BB' = \theta\frac{I}{V}$$

In the present example we have to consider the actual fluid contained in the vessel instead of the imaginary fluid displaced by the vessel but the calculations for the movement of the centre of gravity G are identical with those for the movement of the centre of buoyancy B, so that

$$GG' = \theta\frac{I}{V}$$

(*b*) The stability of the tank can only be tested by giving it a small tilt θ.

If G_0 is the cg of the tank (Fig. 3.11), P the pivot and G the original cg of the liquid which moves to G′ when the tank tilts through an angle θ, then

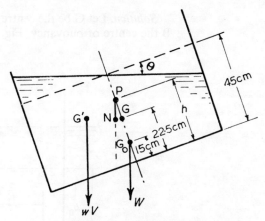

Figure 3.11

Righting moment due to weight of tank $= W \cdot \mathrm{PG}_0 \cdot \theta$

Overturning moment due to contents $= wV \cdot \mathrm{G'N}$

$W = 68 \times 9{\cdot}81 = 668\,\mathrm{N}$, $\mathrm{PG}_0 = h - 0{\cdot}15\,\mathrm{m}$, $w = 9{\cdot}81 \times 10^3\,\mathrm{N/m^3}$

Volume of contents

$$V = 0{\cdot}9 \times 0{\cdot}6 \times 0{\cdot}45 = 0{\cdot}243\,\mathrm{m^3}$$

and G will be 22·5 cm from bottom.

$$\mathrm{G'N} = \mathrm{GG'} - \mathrm{GN} = \mathrm{GG'} - \mathrm{PG} \cdot \theta$$

$$= \theta\frac{I}{V} - \theta(h - 0{\cdot}225)$$

$$I = \frac{0{\cdot}90 \times (0{\cdot}60)^3}{12} = 0{\cdot}01625\,\mathrm{m^4}$$

Since righting moment $=$ overturning moment at the point of instability

$$W \cdot \mathrm{PG}_0 \cdot \theta = wV\left\{\theta\frac{I}{V} - \theta(h - 0{\cdot}225)\right\}$$

$$668(h - 0{\cdot}15) = 9{\cdot}81 \times 10^3 \times 0{\cdot}01625 - 9{\cdot}81 \times 10^3 \times 0{\cdot}243$$
$$\times\,(h - 0{\cdot}225)$$

$$3048h = 795$$

$$h = \mathbf{0{\cdot}26\,m}$$

3.10 Stability of vessel containing liquid

Derive an expression for the effective metacentric height of a vessel which has two longitudinal tanks symmetrically arranged about its axis and containing liquid with a free surface. V is the displacement volume of the vessel, V_1 and V_2 are the volumes of a liquid of specific weight w_1 carried in the tanks, I is the second moment of area of the water plane of the vessel, I_1 and I_2 are the second moments of area of the free surfaces of the liquid in the tanks.

Solution. Let G be the centre of gravity of the vessel and contents and B the centre of buoyancy, Fig. 3.12.

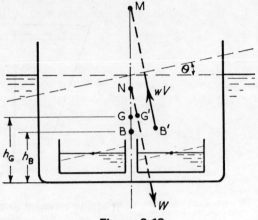

Figure 3.12

When the vessel tilts through a small angle θ the centres of gravity of the liquid in the tanks will move (from Example 3.9) $\theta(I_1/V_1)$ and $\theta(I_2/V_2)$ respectively, causing the centre of gravity of the vessel and contents to move to G'. If w = specific weight of water,

Weight of vessel and contents = weight of water displaced = wV
and taking moments

$$wV \times \text{GG}' = w_1 V_1 \times \theta \frac{I_1}{V_1} + w_2 V_2 \times \theta \frac{I_2}{V_2}$$

$$\text{GG}' = \frac{w_1 \theta (I_1 + I_2)}{wV}$$

The new vertical through G' intersects the original vertical through G at N. The new vertical through B', the displaced position of the centre of buoyancy, cuts the original vertical at the metacentre M.

$$\text{Righting moment} = wV \times \text{NM} . \theta$$

and

Effective metacentric height = NM

$$= h_\text{B} + \text{BM} - (h_\text{G} + \text{GN})$$

$$\text{BM} = \frac{I}{V} \quad \text{and} \quad \text{GN} = \frac{\text{GG}'}{\theta} = \frac{w_1}{w} \frac{(I_1 + I_2)}{V}$$

Effective metacentric height

$$\text{NM} = h_\text{B} - h_\text{G} + \frac{I - \dfrac{w_1}{w}(I_1 + I_2)}{V}$$

Note that the effect of the liquid in the tanks is to reduce the metacentric height and impair stability, but this is only the case if there are free surfaces in the tanks so that the cg of the liquid can move. Subdivision of the tanks improves stability by reducing the sum of the second moments of area of the liquid surfaces.

3.11 Pontoon containing water

A rectangular pontoon 6 m wide, 15 m long and $2 \cdot 1$ m deep weighs 80 metric tons when loaded, but without ballast water. A vertical diaphragm divides the pontoon longitudinally into two compartments each 3 m wide and 15 m long.

Twenty metric tons of water ballast are admitted to the bottom of each compartment, the water surfaces being free to move.

The centre of gravity of the pontoon, without ballast water, is $1 \cdot 5$ m above the bottom and on the geometrical centre of the plan.

(a) Calculate the metacentric height for rolling.

(b) If 2 metric tons of the deck-load is shifted 3 m laterally find the approximate angle of heel.

Solution. Total weight of pontoon and ballast = 120 metric tons = 120 000 kg.

$$\text{Volume of water displaced} = V = 120\,000/1\,000 = 120\,\text{m}^3$$

$$\text{Draught} = 120/6 \times 15 = 1 \cdot 33\,\text{m}$$

$$\text{Height of centre of buoyancy above base} = h_\text{B} = 0 \cdot 667\,\text{m}$$

$$\text{Volume of water ballast} = \frac{2 \times 20 \times 1\,000}{1\,000} = 40\,\text{m}^3$$

$$\text{Depth of ballast water} = \frac{40}{2 \times 3 \times 15} = 0 \cdot 445\,\text{m}$$

$$\text{Height of cg of ballast above bottom} = 0 \cdot 222\,\text{m}$$

Thus, taking moments to find combined cg

Height of cg of vessel and ballast above bottom

$$= h_\text{G} = \frac{80 \times 1 \cdot 5 + 40 \times 0 \cdot 222}{120} = 1 \cdot 074\,\text{m}$$

(a) From Example 3.10

$$\text{Metacentric height} = h_\text{B} - h_\text{G} + \frac{I - \dfrac{w_1}{w}(I_1 + I_2)}{V}$$

$$h_\text{B} = 0 \cdot 667\,\text{m}, \ h_\text{G} = 1 \cdot 074\,\text{m}, \ w_1 = w, \ V = 120\,\text{m}^3$$

$$\text{For the water-line plane } I = 15 \times 6^3/12 = 270\,\text{m}^4$$

$$\text{For each compartment } I_1 = I_2 = 15 \times 3^3/12 = 33 \cdot 8\,\text{m}^4$$

$$\text{Metacentric height} = 0 \cdot 67 - 1 \cdot 074 + \frac{270 - 67 \cdot 6}{120}$$

$$= 1 \cdot 28 \, \text{m for rolling}$$

(b) When the cargo is moved 3 m laterally,

$$\text{Overturning moment} = 2 \times 3 \, \text{tonne-m}$$

$$\text{Righting moment} = W \times \text{NM} \times \theta \qquad \text{(Fig. 3.12)}$$

$$= 120 \times 1 \cdot 28 \times \theta \, \text{tonne-m}$$

$$\text{For equilibrium } 2 \times 10 = 120 \times 1 \cdot 28\theta$$

$$\theta = 0 \cdot 039 \, \text{rad.}$$

$$= 2° \, 14'$$

Problems

1 A rectangular pontoon $5 \cdot 4$ m wide by 12 m long has a draught of $1 \cdot 5$ m in fresh water (density 1000 kg/m³). Calculate (a) the mass of the pontoon, (b) its draught in sea water (density 1025 kg/m³).
Answer 97 200 kg, $1 \cdot 47$ m

2 A vessel lying in a fresh-water dock has a displacement of 10 000 tonnes and the area of the water-line plane is 1840 m². It is moved to a sea-water dock and after removal of cargo its displacement is reduced to 8500 tonnes. Assuming that the sides of the vessel are vertical near the water line and taking the density of fresh water as 1000 kg/m³ and that of sea water as 1025 kg/m³, calculate the alteration in draught.
Answer 0.924 m

3 A cubic metre of ice (sp. gr. $0 \cdot 9$) floats freely in a vessel containing water at 0°C. (a) How much of the ice is exposed, and (b) what will be the change in the level of the water when the ice melts if the area of the water surface is 4 m²?
Answer $0 \cdot 1$ m³, nil

4 The piston of the ball valve shown in Fig. 3.13 has an effective diameter of 10 mm. The valve just closes when $\frac{1}{4}$ of the volume of the ball is submerged. Calculate the pressure in kN/m² of the mains supply.
Answer 552 kN/m²

5 A ball valve is arranged as in Question 4 but the distance from the centre of the ball to the pivot is 300 mm and the diameter of the piston is $12 \cdot 5$ mm. If the mains pressure is 690 kN/m², calculate what fraction of the ball is immersed. What would be the mains pressure if the valve closed when $\frac{1}{3}$ of the ball was immersed?
Answer $0 \cdot 4$, 570 kN/m²

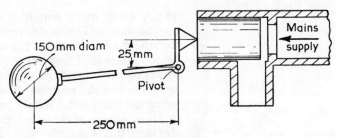

Figure 3.13

6 The sides of a ship are vertical near the waterline and the area of the water-line plane is 2050 m². The total mass of the ship is 10 000 metric tons when it leaves a fresh-water dock. After a certain time at sea the weight of the ship has been reduced by 1500 metric tons due to consumption of fuel. Find how much the draught has been reduced since leaving dock. Relative density of sea water, 1·025.
Answer 0·833 m

7 A rectangular scow 4·5 m by 9·6 m having vertical sides weighs 36 000 kg. What is its draught in fresh water?
Answer 0·835 m

8 A ship has a displacement of 2200 metric tons in sea water. Find the volume of the ship below the water line. Density of sea water 1025 kg/m³.
Answer 2142 m³

9 A ship displaces 13 000 metric tons of sea water. On filling the ship's boats on one side with 60 metric tons of water the angle of heel is 2° 16′. If the boats are 9 m from the centreline of the ship, find the metacentric height. Tan 2° 16′ = 0·0396.
Answer 1·05 m

10 A vessel has a displacement of 1500 metric tons of fresh water. A weight of 16 metric tons moved 8·25 m across the deck causes a horizontal movement of 175 mm at the lower end of a pendulum 1·5 m long. Find the metacentric height.
Answer 0·754 m

11 A rectangular pontoon of mass 90 metric tons floats in sea water. It is 12 m long, 7·5 m wide and 3 m deep. Find the metacentric height. Sea water has a density of 1025 kg/m³ and the centre of gravity of the pontoon may be taken at its geometrical centre.
Answer 3·81 m

12 A rectangular pontoon has a mass of 240 metric tons and a length of 18 m. The centre of gravity is 0·3 m above the centre of cross-section and the metacentric height is to be 1·2 m when the angle of heel is 10 deg. The freeboard must not be less than 0·6 m when the pontoon is vertical. Find the breadth and height of the pontoon if floating in fresh water.
Answer 6·61 m, 2·62 m

13 A vessel has a length of 60 m, a beam of 8·5 m and a displacement of 1350 metric tons of sea water. A mas of 20 metric tons moved 6·75 m across the deck inclines the vessel 5 deg. The second moment of area of the water plane about its fore-and-aft axis is 65 per cent of the second moment of the circumscribing rectangle and the position of the centre of buoyancy is 1·5 m below the water line. Find the position of the metacentre and the centre of gravity of the vessel. Density of sea water, 1025 kg/m³.
Answer 0·074 m above water line, 1·116 m below water line

14 A rectangular pontoon 10·5 m long, 7·2 m broad and 2·4 m deep has a mass of 70 000 kg. It carries on its upper deck a horizontal boiler of 4·8 m diam and a mass 50 000 kg. The centres of gravity of the boiler and the pontoon may be assumed to be at their centres of figure and in the same vertical line. Find the metacentric height. Density of sea water 1025 kg/m³.
Answer 0·865 m

15 State the conditions which govern the stability of a floating vessel.

A buoy carries a light and has a cylindrical upper portion of 2·1 m diam and 1·2 m deep. The lower portion which is curved displaces a volume of 0·396 m³ and its centre of buoyancy is situated 1·28 m below the top of the cylinder. The centre of gravity is situated 1·28 m below the top of the cylinder. The centre of gravity is situated 0·9 m below the top of the cylinder and the total displacement is 2·6 metric tons. Find the metacentric height. Density of sea water is 1025 kg/m³.
Answer 0·316 m

16 A barge is 15 m long and 1·8 m wide and has a flat bottom with vertical sides and ends. The shape in plan is two semicircular ends joined by straight parallel sides. The mass of the barge and its load is 30 metric tons and the centre of gravity is 0·6 m above the bottom. Find the metacentric height when the barge floats in fresh water. If a part of the load having a mass of 1 metric ton is moved from the centre of the barge 0·6 m towards one side, through what angle will the barge turn?
Answer 0·20 m, 5·71 deg

17 A hollow wooden cylinder of sp. gr. 0·55 has an outer diameter of 0·6 m, an inner diameter of 0·3 m and has its ends open. It is required to float in oil of sp. gr. 0·84. Calculate the maximum height of the cylinder so that it shall be stable when floating with its axis vertical and the depth to which it will sink.
Answer 0·53 m, 0·348 m

18 Prove that a circular cylinder floating with its axis horizontal will be in stable equilibrium if its length exceeds the breadth of the water-line section. The proof should cover any depth of immersion.

19 A cylinder has a diameter of 0·3 m and a relative density of 0·8. What is the maximum permissible length in order that it may float with its axis vertical?
Answer 0·266 m

20 Show that a solid cylinder of length L, radius R and specific gravity s will float in stable equilibrium with its axis vertical if $R > \sqrt{[2s(1 - s)]}$.

21 Show that the resultant pressure on the surface of a totally immersed body acts vertically and is equal to the weight of fluid displaced by the body.

A rectangular block of wood, 300 mm by 150 mm in plan, floats immersed to a depth of 75 mm. Find the height of the longitudinal and transverse metacentres above its bottom.
Answer 62·5 mm, 137·5 mm

22 A solid cube of wood of specific gravity 0·9 floats in water with a face parallel to the water plane. If the length of one edge is 10 cm, find the metacentric height.
Answer 0·0043 m

23 Derive the expression $2\pi\sqrt{(K^2/gh)}$ for the period of oscillation of a floating body about a position of stability, assuming that it oscillates about its cg. K is the relevant radius of gyration and h is the metacentric height.

A pontoon 6 m × 2·4 m × 1·2 m deep, which may be considered to be a hollow rectangular box with walls 12·5 mm thick, is made of steel weighing 7750 kg/m³. Calculate the period of oscillation about the longer axis when floating in fresh water. Secondary effects of thickness may be disregarded.
Answer 1·88 sec

24 Show that a buoy 1·8 m in diam and 2·4 m high with a mass of 1800 kg will not float with its axis vertical in sea water (density 1025 kg/m³). What pull must be applied to a vertical chain fastened to the centre of the base so that the buoy will just float with its axis vertical?
Answer 11·7 kN

25 A solid buoy made of material 0·6 times as dense as sea water floats in sea water. The buoy consists of an upright cylinder 1·2 m in diam and 1·5 m long with the addition at the lower end of a hemisphere 1·2 m in diam of the same material. A chain is attached to the lowest point of the hemisphere.

Find the required vertical pull on the chain so that the buoy just floats with the axis of the cylindrical portion vertical.

The centroid of a hemisphere of radius r is $\frac{3}{8}r$ from the centre of curvature. Sea water has a density of 1025 kg/m³.
Answer 2·71 kN

26 A ship displaces 10 000 metric tons and the area of its plane of flotation is 1480 m². The centre of mass is 49 m and the centre of area of the plane of flotation is 55 m from the stern. The metacentric height for pitching motion about the transverse principal axis is 91·5 m. The ship is loaded in sea water with 300 metric tons of extra cargo. Find the minimum allowable distance of the mass centre of this extra load from the stern if, when the ship passes from sea water into a freshwater canal, the stern draught must not increase by more than 0·3 m. Assume that the

metacentric height and the area of the plane of flotation are not altered by the change in draught and that the density of sea water is 1025 kg/m³.
Answer   46·2 m

27   A ship is inclined at a small angle from the vertical. Show, by means of a sketch, the forces acting on the ship, and explain what is meant by the "righting moment."

A raft is formed of 3 cylinders, each 1·2 m in diam, symmetrically placed with their axes horizontal, the extreme breadth over the cylinders being 6 m. The fore-and-aft axis is parallel to the axes of the cylinders. When laden the raft floats with the cylinders half immersed and its centre of gravity in this condition is 1·2 m above the centre cylinder axis. Calculate the transverse metacentric height.
Answer   6·95 m

28   The half-ordinates of the contour of the load water plane of a ship of 2800 metric tons displacement, measured at intervals of 6 m, are 0·30, 3·33, 5·07, 6·18, 6·75, 6·90, 6·75, 6·18, 5·61, 4·47, 0·60 m respectively. If the centre of gravity is 1·87 m above the centre of buoyancy, calculate the transverse metacentric height. Sea water has a density of 1025 kg/m³. Explain the relation between the locus of the metacentre and the curve of buoyancy as the vessel heels over.
Answer   0·74 m

29   A hollow cylinder with closed ends is 300 mm diam and 450 mm high, has a mass of 27 kg and has a small hole in the bottom. It is lowered into water so that its axis remains vertical. Calculate the depth to which it will sink, the height to which the water will rise in it and the air pressure inside it. Disregard the effect of the thickness of the walls but assume that it is uniform and that the compression of the air is isothermal. Atmospheric pressure is 1·02 bar.

Determine also whether the cylinder will be stable in the vertical position when in equilibrium.
Answer   0·398 m 15·85 mm, 3750 N/m², unstable

30   A rectangular pontoon 9 m wide, 12 m long and 4.5 m deep is divided longitudinally by a vertical bulkhead into two equal compartments partly filled with water ballast, the surface of which is free to move. If the total displacement of the pontoon is 200 metric tons, find the height of its cg above the bottom if the metacentric height is 0·6 m.
Answer   3·06 m

31   A tank barge of unladen weight 1000 kN is 24 m long, 4·5 m wide, and the cg of the structure is 1·2 m above the keel. It is divided into two tanks, each 18 m long and 1·8 m wide, the base of the tanks being 0·3 m above the keel. One tank is partially filled with 45 000 litres of oil of sp. gr. 0·89, the other tank is empty. The draught under these conditions is 2·1 m and the centre of buoyancy is then 0·9 m below the surface. To what

angle will the barge tilt? Take the moment of inertia of the water-line plane as 75 per cent of the circumscribing rectangle.
Answer 16°

32 A hollow cylinder 2·4 m high consists of two concentric cylindrical shells, each 6 mm thick, of 2·4 m and 3·0 m diam respectively. The top is open and the bottom is totally closed by a plate 25 mm thick. The cylinder floats with its axis vertical. Determine the metacentric height. The density of the material is 7700 kg/m³. Find also the new metacentric height when there is water in the inner chamber of 2·4 m diam to a depth of 0·6 m.
Answer 0·725 m, 0·298 m

33 A rectangular pontoon is 12 m long, 4·2 m broad and 2·1 m deep, and it floats in fresh water. The pontoon has a displacement of 65 metric tons unloaded and the centre of gravity is 0·6 m up from the bottom and 20 mm aside from the vertical centreline. Determine the inclination of the centreline when the pontoon floats freely.

If the pontoon becomes partially flooded, find the inclination of the centreline when the depth of water inside the pontoon at the centre is 200 mm.
Answer 58′, 4° 40′

4

Liquids in relative equilibrium

If a vessel containing a liquid is at rest or moving with constant linear velocity the liquid is *not* affected by the motion of the container, but if the container is given a continuous acceleration this will be imparted to the liquid which will take up a new position and come to rest relative to the vessel. The liquid is in relative equilibrium and is at rest with respect to its container. There is no relative motion of the particles of the fluid and therefore no shear stress. Fluid pressure is everywhere normal to the surface on which it acts.

4.1 Horizontal acceleration

A tank containing water moves horizontally with a constant linear acceleration f of 3 m/s². The tank is 3 m long and the depth of water when the tank is at rest is 1·5 m. Calculate (*a*) the angle of the water surface to the horizontal, (*b*) the maximum pressure intensity on the bottom, and (*c*) the minimum pressure intensity on the bottom.

Solution. (*a*) Consider a particle O of mass m on the free surface of the liquid (Fig. 4.1). Since the particle is at rest relative to the tank, it

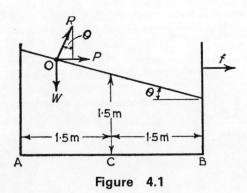

Figure   4.1

will have the same acceleration f, and will be subjected to an accelerating force P

$$P = mf = \frac{W}{g} f$$

where W = weight of particle.

The accelerating force P is the resultant of the weight W of the particle acting vertically downward and the pressure force R acting normal to the free surface due to the surrounding fluid. For equilibrium $P = W \tan \theta$ where θ is the angle of the free surface to the horizontal.

Thus
$$W \tan \theta = P = W\frac{f}{g}$$

$$\tan \theta = \frac{f}{g}$$

and is constant for all points on the free surface.

$$\text{Putting } f = 3\,\text{m/s}^2, \tan \theta = \frac{3}{9 \cdot 81} = 0 \cdot 306$$

$$\text{Angle of water surface to horizontal} = \mathbf{17°}$$

(b) Since the accelerating force is horizontal, vertical forces are not affected. The intensity of pressure at any depth h will be wh where w is the specific weight of the liquid and equals ρg.

Maximum pressure intensity occurs at A where the depth is h_A. Since the contents of the tank are unchanged

$$\text{Depth at centre C} = 1 \cdot 5\,\text{m}$$

$$h_A = 1 \cdot 5 + 1 \cdot 5 \tan \theta = 1 \cdot 5 + 0 \cdot 46 = 1 \cdot 96\,\text{m}$$

Pressure intensity at A $= \rho g h_A = 10^3 \times 9 \cdot 81 \times 1 \cdot 96 = 19 \cdot 2\,\text{kN/m}^2$

(c) Minimum pressure occurs at B

$$\text{Depth at B} = h_B = 1 \cdot 5 - 1 \cdot 5 \tan \theta = 1 \cdot 04\,\text{m}$$

Pressure intensity at B $= \rho g h_B = 10^3 \times 9 \cdot 81 \times 1 \cdot 04 = 10 \cdot 2\,\text{kN/m}^2$

4.2 Vertical acceleration

A vessel containing a liquid of mass density 930 kg/m³ is given a constant vertical acceleration $f = 4 \cdot 8$ m/s² in an upward direction. If the vessel is $1 \cdot 2$ m wide and $1 \cdot 5$ m in breadth and the depth of the liquid is $0 \cdot 9$ m, calculate the force on the bottom of the vessel (a) while it is being accelerated, (b) when the acceleration ceases and the vessel continues to move at a constant velocity of 6 m/s vertically upward.

Solution. As the acceleration is vertical the free surface will remain horizontal. Consider a vertical prism of height h (Fig. 4.2) extending from the free surface to X and let the pressure intensity at X be p.

Accelerating force at X $= P$

$$= \text{force due to pressure} - \text{weight of prism}$$

$$= pa - \rho g h a$$

By Newton's 2nd Law $P = \text{mass} \times \text{acceleration}$

$$= \rho h a \times f$$

Thus $\qquad pa - \rho gha = \rho ha \times f$

$$p = \rho gh\left(1 + \frac{f}{g}\right) \qquad . \qquad . \qquad . \qquad . \qquad . \qquad (1)$$

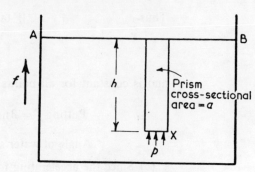

Figure 4.2

(a) From equation (1),

$$p = \rho gh\left(1 + \frac{f}{g}\right)$$

$\rho g = 930 \times 9\cdot81\,\text{N/m}^3$, $h = 0\cdot9\,\text{m}$, $f = 4\cdot8\,\text{m/s}^2$

Intensity of pressure $= p = 930 \times 9\cdot81 \times 0\cdot9\left(1 + \dfrac{4\cdot8}{9\cdot81}\right)$

$$= 12\,200\,\text{N/m}^2$$

Area of base of tank $= 1\cdot2 \times 1\cdot5 = 1\cdot8\,\text{m}^2$

Force on base of tank $= 12\,200 \times 1\cdot8 = \mathbf{22\,000\,N}$

(b) When the velocity is uniform $f = 0$ and

$$p = \rho gh = 930 \times 9\cdot81 \times 0\cdot9 = 8200\,\text{N/m}^2$$

Force on base of tank $= 8200 \times 1\cdot8 = \mathbf{14\,800\,N}$

4.3 Forced vortex

A cylindrical tank is spun at 300 rev/min with its axis vertical. The tank is $0\cdot6$ m high and 45 cm diam and is filled completely with water before spinning. Show that the water surface will take the form of a paraboloid when the container is spun and calculate (a) the speed at which the water surface will just touch the top rim and centre bottom of the tank, (b) the level to which the water will return when the tank stops spinning and the amount of water lost.

Solution. The liquid in the vessel will eventually rotate with the vessel at the same angular velocity ω. A particle on the free surface will be in equilibrium under the action of its weight W (Fig. 4.3), the centrifugal accelerating force P acting horizontally and the fluid reaction R. For any point at radius x and a height y from the lowest point O, if θ is the

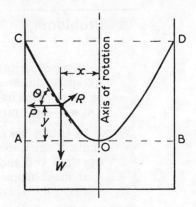

Figure 4.3

angle of inclination of the water surface to the horizontal,

$$\tan \theta = \frac{dy}{dx} = \frac{P}{W}$$

as in Example 4.1.

For a constant value of ω, P will vary with x, since the centrifugal acceleration is $\omega^2 x$ and $P = (W/g)\omega^2 x$. The surface angle therefore varies and

$$\tan \theta = \frac{dy}{dx} = \frac{\omega^2 x}{g}$$

Integrating

$$y = \int_0^z \frac{\omega^2 x}{g}\, dx = \frac{\omega^2 x^2}{2g} + \text{constant}$$

If y is measured from AB, $y = 0$ when $x = 0$ and

$$y = \frac{\omega^2 x^2}{2g}$$

The water surface is therefore a paraboloid of revolution:

(a) When the surface just touches the rim and the base, $y = 0.6\,\text{m}$ and $x = 0.225\,\text{m}$

$$\omega = \sqrt{\frac{2gy}{x^2}} = \sqrt{\frac{2g \times 0.6}{(0.225)^2}}\ \text{rad/s}$$

$$= 15.15\,\text{rad/s}$$

$$= \mathbf{144.5\,rev/s}$$

(b) In Fig. 4.3 the volume of the paraboloid will be half that of the containing cylinder ABDC. When the water surfaces touch the rim and the centre of the bottom,

$$\text{Volume of water left in tank} = \text{half of original volume}$$

$$= \tfrac{1}{2}\pi \times (0.225)^2 \times 0.6$$

$$= 0.0475\,\text{m}^3$$

$$\text{Depth when tank ceases to rotate} = 0.3\,\text{m}$$

$$\text{Amount of water thrown out} = \mathbf{0.0475\,m^3}$$

Problems

1 A vessel partly filled with liquid and moving horizontally with a constant linear acceleration has its liquid surface inclined at 45°. Determine its acceleration.
Answer 9·81 m/sec²

2 A U-tube has a horizontal part 0·6 m long with vertical end limbs. If the whole tube is rotated about a vertical axis 0·45 m from one end and 0·15 m from the other, calculate the speed when the difference of level between the tubes is 0·25 m
Answer 49·9 rev/min

3 A cylindrical vessel, 100 mm in diam and 0·3 m high, contains water when at rest to a depth of 225 mm. If the vessel is rotated about its longitudinal axis, which is vertical, calculate from first principles the speed at which the water will commence to spill over the edge, and the speed when the axial depth is zero.
Answer 34·4 rad/s, 48·6 rad/s

4 A glass tube, internal diam 50 mm and length 300 mm, has its axis vertical. It is closed at both ends and contains a liquid filling three-fourths of the volume of the tube. The tube is made to revolve about its axis. Find the speed in rev/min when the bottom of the cup formed by the liquid is at the bottom of the tube. This arrangement is used as a speed indicator. Plot a graph showing the relation between speed in rev/min and the depth of the vortex below the top of the tube.
Answer 1311 rev/min

5 A flat cylindrical disc of 0·45 m diam, keyed on to the lower end of a 75 mm diam shaft, serves as an hydraulic footstep bearing. Its lower face is plane and bears against radiating ribs cast in the pressure cylinder, while its upper face carries a series of radial ribs which bear against the plane upper lid of the pressure cylinder. Water above rotates with the disc, that below is stationary and the upper and lower sides are in free communication round the edge of the disc. Determine the resulting pressure on the shaft at 400 rev/min.
Answer 3530 N

6 What is the greatest speed in revolutions per min at which an open cylindrical tank 0·6 m in diam may be rotated about its vertical axis, the tank being 0·9 m high and two-thirds filled with water when stationary, if (*a*) no water is to spill over the sides, (*b*) the water spills over the sides and the bottom of the tank is free of water for a radius of 150 mm about the vertical axis?
Answer 109 rev/min, 154·5 rev/min

7 A tube AOB has the part OB, 300 mm long, bent upwards so that the angle AOB is 45°. AO is vertical and the end B is closed. The tube AOB is completely filled with water to a height of 230 mm above O. Find the number of revolutions per min of the tube about the axis OA so that the pressure at B is the same as the pressure at O. What is the least pressure in OB and where does this occur?

Answer 91·9 rev/min, 177 mm, 150 mm from O

8 An hydraulic footsteps bearing on a 100 mm diam shaft consists of a 500 mm diam disc rotating in water in a pressure cylinder. The lower face of the disc is plane and is supported by radial ribs on the base of the pressure cylinder, its upper face has radial ribs and the upper surface of the pressure cylinder is plane. The two portions of the cylinder are in free communication with each other around the periphery of the disc. Find the load which the bearing will support at 500 rev/min without pressure between the disc and the fixed ribs.

Answer 8400 N

9 A cylindrical bucket containing water is swung in a vertical plane so that the bottom of the bucket describes a circle 2 m in diam at constant speed. The bucket is 350 mm deep and 300 mm diam and contains water to a depth of 300 mm. Calculate the minimum speed of rotation if no water is to escape from the bucket.

Answer 35·75 rev/min

10 A conical vessel, with base uppermost, is rotated about its axis which is vertical. The vessel was completely filled with water when at rest. After rotating the vessel at 60 rev/min only 150 litres of water remain in it. Calculate the height of the cone if the diameter of the base is 1 metre.

Answer 1·31 m

5

Liquids in motion

There is no shear force in a fluid at rest but, when it is in motion, shear forces can be set up due to viscosity and turbulence which oppose motion, producing a "frictional" effect. Many problems can be solved, at least partially, by assuming an ideal frictionless (inviscid) fluid. Where necessary an experimental coefficient is introduced to make the theoretical results agree with the observed values and to represent the effect of the factors such as frictional resistance omitted from the theory.

A fluid consists of a large number of individual particles moving in the general direction of flow but usually not parallel to each other. The velocity of any particle is a vector quantity having magnitude and direction which vary from moment to moment. The path followed by a particle is called a *path line*. At any given instant of time the positions of successive particles can be joined up by a curve which is tangential to the

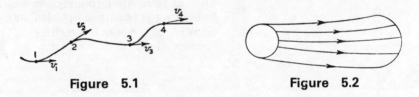

Figure 5.1   Figure 5.2

direction of motion of the particle at that instant (Fig. 5.1). This curve is called a *streamline* and is ordinarily a curve in three dimensions. Since the velocity of a particle at any point on a streamline is tangential to it, there can be no flow across a streamline.

When considering the flow of a large body of fluid it is sometimes convenient to consider a small section. If streamlines are drawn through every point on the circumference of a small area (Fig. 5.2) they form a *stream tube*. Since there is no flow across a streamline the fluid inside a stream tube cannot escape and can be treated conveniently.

Types of flow

Two distinct types of flow occur which are

1. *Turbulent flow*, in which the particles of fluid move in a disorderly manner, occupying different relative positions in successive cross-sections.

2. *Viscous flow*, also known as *streamline* or *laminar* flow, in which the particles of the fluid move in an orderly manner and retain the same relative positions in successive cross-sections.

Osborne Reynolds found that the type of flow is determined by the velocity, density and viscosity of the fluid and the size of the conduit and depends on the value of $\rho v d/\mu$ (Reynolds number) in which v = velocity, ρ = mass density and μ = viscosity of the fluid, while d is a typical dimension which for a pipe is the diameter.

For flow in pipes if the Reynolds number is less than 2100 approximately, flow is always viscous.

5.1 Definitions

Define the following terms used in connexion with the flow of a liquid: (*a*) uniform flow, (*b*) steady flow, (*c*) unsteady flow, (*d*) mean velocity, (*e*) discharge, (*f*) mass flow rate.

Solution. (*a*) *Uniform flow.* The cross-sectional area and velocity of the stream of fluid are the same at each successive cross-section. Example: flow through a pipe of uniform bore running completely full.

(*b*) *Steady flow.* The cross-sectional area and velocity of the stream may vary from cross-section to cross-section, but for each cross-section they do not change with time. Example: flow through a tapering pipe.

(*c*) *Unsteady flow.* The cross-sectional area and velocity of the stream at any cross-section vary with time. Example: a wave travelling along a channel.

(*d*) *Mean velocity.* The mean velocity V at any cross-section of area A when the volume passing per second is Q, will be

$$V = \frac{Q}{A}$$

(*e*) *Discharge.* The volume of liquid passing a given cross-section in unit time is called the discharge. It is measured in cubic metres per second, or similar units, and denoted by Q.

(*f*) *Mass flow rate.* The mass of fluid passing a given cross-section in unit time is called the mass flow rate. It is measured in kilogrammes per second, or similar units and denoted by $\dot{m}$.

5.2 Continuity of flow

What is meant by continuity of flow and under what conditions does it occur?

Oil flows through a pipe line (Fig. 5.3) which contracts from 450 mm diam at A to 300 mm diam at B and then forks, one branch being 150 mm diam discharging at C and the other branch 225 mm diam discharging at D. If the velocity at A is 1·8 m/s and the velocity at D is 3·6 m/s, what will be the discharges at C and D and the velocities at B and C?

Solution. For continuity of flow in any system of fluid flow the total amount of fluid entering the system must equal the amount leaving the system. This occurs in the case of uniform flow and of steady flow.

In pipe AB (Fig. 5.3) if Q_A is the discharge at A and Q_B the discharge at B,

$$Q_A = Q_B \text{ or } a_A v_A = a_B v_B$$

where a_A and a_B are the cross-sectional areas at A and B. Similarly, if Q_C and Q_D are the discharges at C and D

$$Q_A = Q_C + Q_D$$
$$a_A v_A = a_C v_C + a_D v_D$$

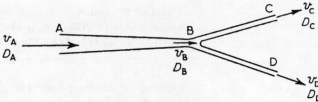

Figure 5.3

If the continuity of flow relation did not apply fluid would have to be created or destroyed within the system, otherwise the pipe line would collapse or burst.

For compressible fluids there is continuity of flow when the mass passing each section, or entering and leaving the system, is the same.

$$\text{Discharge at A} = Q_A = \tfrac{1}{4}\pi D_A^2 v_A$$
$$= \tfrac{1}{4}\pi \times (0.450)^2 \times 1.8 = 0.287 \,\text{m}^3/\text{s}$$
$$\text{Discharge at C} = Q_C = \tfrac{1}{4}\pi D_C^2 v_C$$
$$= \tfrac{1}{4}\pi (0.150)^2 v_C = 0.0177 v_C \,\text{m}^3/\text{s}$$
$$\text{Discharge at D} = Q_D = \tfrac{1}{4}\pi D_D^2 v_D$$
$$= \tfrac{1}{4}\pi (0.225)^2 \times 3.6 = 0.143 \,\text{m}^3/\text{s}$$

For continuity of flow

$$Q_A = Q_C + Q_D$$
$$\therefore \qquad Q_C = Q_A - Q_D = 0.287 - 0.143 = 0.144 \,\text{m}^3/\text{s}$$

Also for continuity of flow between A and B,

$$Q_A = Q_B, \quad \tfrac{1}{4}\pi D_A^2 v_A = \tfrac{1}{4}\pi D_B^2 v_B$$

$$v_B = v_A \left(\frac{D_A}{D_B}\right)^2 = 1.8 \left(\frac{0.450}{0.300}\right)^2 = 4.05 \,\text{m/s}$$

Since $Q_C = 0.144 \,\text{m}^3/\text{s}$

$$v_C = \frac{Q_C}{\tfrac{1}{4}\pi D_C^2} = \frac{0.144}{\tfrac{1}{4}\pi (0.150)^2} = 8.1 \,\text{m/s}$$

Rate of change of momentum

5.3 Momentum of a fluid

A fluid is flowing in a tapering pipe (Fig. 5.4). At section AB the area of cross-section is a_1 and the velocity v_1 and at section CD the corresponding values are a_2 and v_2. Derive an expression for the rate of change of momentum of the fluid between the two sections.

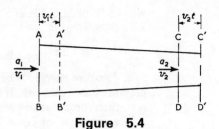

Figure 5.4

Solution. Suppose that in a small time interval t the fluid contained between AB and CD moves to A'B' and C'D'.

Change of momentum of this fluid in time $t =$ Increase due to momentum of fluid CC'D'D $-$ momentum of fluid AA'B'B

$$\text{Distance } CC' = v_2 t$$

$$\text{Mass of fluid } CC'D'D = \rho a_2 v_2 t$$

where $\rho =$ mass density of fluid

$$\text{Momentum of fluid } CC'D'D = \rho a_2 v_2 t \times v_2 = \rho a_2 v_2{}^2 t$$

Similarly, $\qquad$ Momentum of fluid AA'B'B $= \rho a_1 v_1{}^2 t$

Change of momentum of fluid between AB and CD in time t

$$= \rho a_2 v_2{}^2 t - \rho a_1 v_1{}^2 t$$

Rate of change of momentum of fluid between AB and CD

$$= \rho(a_2 v_2{}^2 - a_1 v_1{}^2)$$

For continuity of flow, $\quad a_1 v_1 = a_2 v_2 = Q$

where Q is the discharge.

Rate of change of momentum of fluid between AB and CD

$$= \rho Q(v_2 - v_1) = \frac{wQ}{g}(v_2 - v_1)$$

$$= \text{mass per sec flowing} \times \text{change of velocity}$$

Energy of a flowing liquid

5.4 Energy of a fluid

State the forms of energy which a liquid in motion can possess and derive expressions for each of these forms in terms of the pressure p, velocity v and elevation z for unit weight of fluid.

Water at an altitude of 36 m above sea level has a velocity of 18 m/s and a pressure of 350 kN/m². Calculate the total energy per newton of this water reckoned above sea level.

Solution. A liquid may possess three forms of energy:

1. *Potential energy* because of its elevation above datum level. If a

weight W of liquid is at a height of z above datum

$$\text{Potential energy} = Wz$$

∴ $\quad\quad\quad$ Potential energy per unit weight $= z$

2. *Pressure energy.* When a fluid flows in a continuous stream under pressure it can do work. If the area of cross-section of the stream of fluid is a, then force due to pressure p on cross-section is pa.

If a weight W of liquid passes the cross-section

$$\text{Volume passing cross-section} = \frac{W}{w}$$

$$\text{Distance moved by liquid} = \frac{W}{wa}$$

$$\text{Work done} = \text{force} \times \text{distance} = pa \times \frac{W}{wa}$$

$$= W\frac{p}{w}$$

$$\text{Pressure energy per unit weight} = \frac{p}{w} = \frac{p}{\rho g}$$

Note "pressure energy" is the energy of a fluid *flowing* under pressure. Do not confuse it with the energy stored in a fluid, due to its elasticity, when it is compressed.

3. *Kinetic energy.* If a weight W of liquid has a velocity v,

$$\text{Kinetic energy} = \tfrac{1}{2}\frac{W}{g}v^2$$

$$\text{Kinetic energy per unit weight} = v^2/2g$$

The total energy of the liquid is the sum of these three forms of energy.

$$\text{Total energy per unit weight} = z + \frac{p}{w} + \frac{v^2}{2g}$$

Putting $z = 36\,\text{m}$, $p = 350\,\text{kN/m}^2 = 350 \times 10^3\,\text{N/m}^2$, $v = 18\,\text{m/s}$ and $w = 9\cdot81 \times 10^3\,\text{N/m}^3$

$$\text{Total energy per unit weight} = 36 + \frac{350 \times 10^3}{9\cdot81 \times 10^3} + \frac{18^2}{2 \times 9\cdot81}$$

$$= 36 + 35\cdot7 + 16\cdot5 = \mathbf{88\cdot2\,N\text{-}m/N}$$

5.5 Bernoulli's equation for frictionless flow

What is meant by (a) potential head, (b) pressure head, (c) velocity head, (d) total head for a liquid in motion?

State Bernoulli's theorem for a liquid.

A jet of water from a 25 mm diam nozzle is directed vertically upwards. Assuming that the jet remains circular and neglecting any loss of energy, what will be the diameter of the jet at a point $4\cdot5$ m above the nozzle if the velocity with which the jet leaves the nozzle is 12 m/s.

Solution. (*a*) The potential head is another term for the potential energy per unit weight. Referring to Example 5.4 the potential energy per unit weight has dimensions of N-m/N and is measured as a length or head z and can be called the *potential head*.

(*b*) Similarly the pressure energy per unit weight p/w is equivalent to a head and is referred to as the *pressure head*.

(*c*) The kinetic energy per unit weight $v^2/2g$ is also measured as a length and referred to as the *velocity head*.

Total head = potential head + pressure head + velocity head

$$H = z + \frac{p}{w} + \frac{v^2}{2g}$$

Bernoulli's theorem states that the total energy of each particle of a body of fluid is the same provided that no energy enters or leaves the system at any point. The division of this energy between potential, pressure and kinetic energy may vary, but the total remains constant. In symbols:

$$H = z + \frac{p}{w} + \frac{v^2}{2g} = \text{constant}$$

Referring to Fig. 5.5, let v_1, d_1, and z_1 be the velocity and diameter of the jet and the elevation of the nozzle respectively, and let v_2, d_2 and z_2 be the corresponding values at the upper level. At both sections the water is at atmospheric pressure p_A.

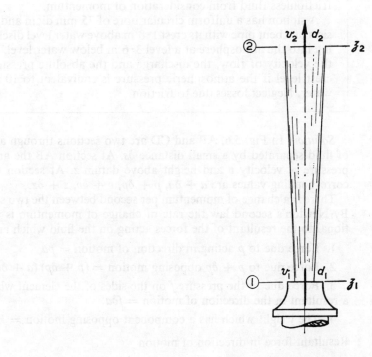

Figure 5.5

By Bernoulli's theorem,

Total energy per unit weight at section 1

$$= \text{total energy per unit weight at section 2}$$

$$z_1 + \frac{p_A}{w} + \frac{v_1^2}{2g} = z_2 + \frac{p_A}{w} + \frac{v_2^2}{2g}$$

$$\frac{v_1^2 - v_2^2}{2g} = z_2 - z_1$$

Putting $v_1 = 12\,\text{m/s}$ and $z_2 - z_1 = 4\cdot5\,\text{m}$

$$12^2 - v_2^2 = 2g \times 4\cdot5$$

$$v_2^2 = 144 - 88\cdot3 = 55\cdot7$$

$$v_2 = 7\cdot46\,\text{m/s}$$

For continuity of flow

$$\tfrac{1}{4}\pi d_1^2 v_1 = \tfrac{1}{4}\pi d_2^2 v_2$$

$$d_2 = d_1 \sqrt{\frac{v_1}{v_2}} = 25 \sqrt{\frac{12}{7\cdot46}} = \mathbf{31\cdot7\,mm}$$

5.6 Bernoulli's equation for frictionless flow

> Derive Bernoulli's equation for the flow of an incompressible frictionless fluid from consideration of momentum.
>
> A siphon has a uniform circular bore of 75 mm diam and consists of a bent pipe with its crest $1\cdot8$ m above water level discharging into the atmosphere at a level $3\cdot6$ m below water level. Find the velocity of flow, the discharge and the absolute pressure at crest level if the atmospheric pressure is equivalent to 10 m of water. Neglect losses due to friction.

Solution. In Fig. 5.6, AB and CD are two sections through a stream of fluid separated by a small distance δs. At section AB the area is a, pressure p, velocity v and height above datum z. At section CD the corresponding values are $a + \delta a$, $p + \delta p$, $v + \delta v$, $z + \delta z$.

There is a change of momentum per second between the two sections. By Newton's second law the rate of change of momentum is proportional to the resultant of the forces acting on the fluid which are

1. Force due to p acting in direction of motion $= pa$
2. Force due to $p + \delta p$ opposing motion $= (p + \delta p)(a + \delta a)$
3. Force due to the pressure f on the sides of the element which has a resultant in the direction of motion $= f\delta a$
4. The weight which has a component opposing motion $= W \cos \theta$

Resultant force in direction of motion

$$= pa - (p + \delta p)(a + \delta a) + f\delta a - W \cos \theta$$

The value of f must vary from p at AB to $p + \delta p$ at CD and can be taken as $p + k\delta p$ where k is a fraction.

$$W = \text{spec. weight} \times \text{volume of element} = w(a + \tfrac{1}{2}\delta a)\delta s$$

$$\cos \theta = \frac{\delta z}{\delta s}$$

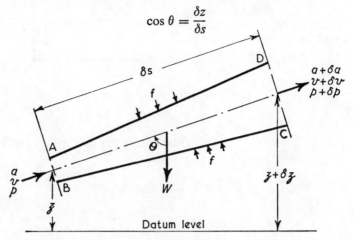

Figure 5.6

Force in direction of motion

$$= -p\delta a - a\delta p - \delta p \,.\, \delta a + p\delta a + k\delta p \,.\, \delta a - w(a + \tfrac{1}{2}\delta a)\delta s \,.\, \frac{\delta z}{\delta s}$$
$$= -a\delta p - wa\delta z$$

neglecting products of small quantities.

Rate of change of momentum of fluid

$$= \text{mass/sec} \times \text{change of velocity} = \frac{wav}{g} \,.\, \delta v$$

Rate of change of momentum = Applied force

$$\frac{wav}{g} \,.\, \delta v = -a\delta p - wa\delta z$$

$$\delta z + \frac{v\delta v}{g} + \frac{\delta p}{w} = 0 \quad . \quad . \quad . \quad . \quad (1)$$

For an incompressible fluid w is constant.

Integrating along the stream

$$z + \frac{v^2}{2g} + \frac{p}{w} = \text{constant}$$

or between any two points,

$$z_1 + \frac{v_1{}^2}{2g} + \frac{p_1}{w} = z_2 + \frac{v_2{}^2}{2g} + \frac{p_2}{w}$$

This is Bernoulli's equation for an incompressible frictionless fluid.

Referring to Fig. 5.7, apply Bernoulli's equation to points A and C, taking C as datum level, atmospheric pressure as p_0 and velocity at A as zero.

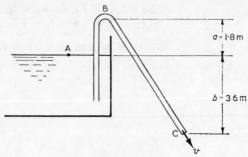

Figure 5.7

Total energy per unit wt at A = Total energy per unit wt at C

$$b + 0 + \frac{p_0}{w} = 0 + \frac{v^2}{2g} + \frac{p_0}{w}$$

$$\frac{v^2}{2g} = b = 3 \cdot 6 \, \text{m}$$

Velocity of flow $v = \sqrt{(2g \times 3 \cdot 6)} = \textbf{8·4m/s}$

Discharge = area × velocity = $\frac{1}{4}\pi(0 \cdot 075)^2 \times 8 \cdot 4 = \textbf{0·0371 m}^3\textbf{/s}$

Applying Bernoulli's equation to A and to the crest B where the absolute pressure is p_B and velocity v

Total energy per unit wt at A = Total energy per unit wt at B

$$b + 0 + \frac{p_0}{w} = (a + b) + \frac{v^2}{2g} + \frac{p_B}{w}$$

Putting $p_0/w = 10$ m of water, $v = 8 \cdot 4$ m/s,

$$3 \cdot 6 + 10 = 5 \cdot 4 + \frac{8 \cdot 4^2}{2g} + \frac{p_B}{w}$$

$$\frac{p_B}{w} = 13 \cdot 6 - 9 \cdot 0 = 4 \cdot 6 \, \text{m of water}$$

Absolute pressure at B = $4 \cdot 6 \times 9 \cdot 81 \times 10^3 = \textbf{45·1} \times \textbf{10}^3 \textbf{N/m}^2$

5.7 Loss of energy

Explain how provision can be made in Bernoulli's equation for a loss of energy occurring between two points in a stream of liquid.

A conical tube is fixed vertically with its smaller end upwards. The velocity of flow down the tube is 4·5 m/s at the upper end and 1·5 m/s at the lower end. The tube is 1·5 m long and the pressure head at the upper end is 3 m of the liquid. The loss in the tube expressed as a head is $0 \cdot 3(v_1 - v_2)^2/2g$ where v_1 and v_2 are the velocities at the upper and lower end. What is the pressure head at the lower end?

Solution. The loss of energy between the two points A and B in the fluid is expressed as a loss of energy per unit weight flowing (e.g. in N-m/N), that is to say as a loss of head. Bernoulli's equation for flow from A to B can then be written for a unit weight as

Total energy at A = Total energy at B + Loss of energy from A to B

$$z_A + \frac{p_A}{w} + \frac{v_A{}^2}{2g} = z_B + \frac{p_B}{w} + \frac{v_B{}^2}{2g} + \text{Loss of head}$$

Let p_1, v_1 and z_1 be the pressure, velocity and height above datum at the upper end of the tube and p_2, v_2 and z_2 the corresponding values at the lower end.

Total energy at upper end = Total energy at lower end + Loss

$$\frac{p_1}{w} + z_1 + \frac{v_1{}^2}{2g} = \frac{p_2}{w} + z_2 + \frac{v_2{}^2}{2g} + \frac{0 \cdot 3(v_1 - v_2)^2}{2g}$$

$$\frac{p_2}{w} = \frac{p_1}{w} + (z_1 - z_2) + \frac{v_1{}^2}{2g} - \frac{v_2{}^2}{2g} - \frac{0 \cdot 3(v_1 - v_2)^2}{2g}$$

$$\frac{p_1}{w} = 3\,\text{m}, \ (z_1 - z_2) = 1 \cdot 5\,\text{m}, \ v_1 = 4 \cdot 5\,\text{m/s}, \ v_2 = 1 \cdot 5\,\text{m/s}$$

$$\frac{p_2}{w} = 3 + 1 \cdot 5 + \frac{4 \cdot 5^2}{2g} - \frac{1 \cdot 5^2}{2g} - \frac{0 \cdot 3(4 \cdot 5 - 1 \cdot 5)^2}{2g}$$

Pressure head at lower end $= \dfrac{p_2}{w} = \mathbf{5 \cdot 28\,m\ of\ liquid}$

5.8 Power of a jet

A jet of water is discharged through a nozzle with an effective diameter d of 75 mm and a velocity v of $22 \cdot 5$ m/s. Calculate the power of the issuing jet.

If the nozzle is supplied from a reservoir which is 30 m above it, what is the loss of head in the pipe line and nozzle and the efficiency of power transmission?

Solution.

$$\text{Kinetic energy per unit wt of jet} = \frac{v^2}{2g}$$

If W = weight issuing per second

$$\text{power of jet} = \frac{Wv^2}{2g}$$

$$W = wQ = wav$$

$$\text{power of jet} = \frac{wav^3}{2g}$$

$$a = \tfrac{1}{4}\pi d^2 = \tfrac{1}{4}\pi(0 \cdot 075)^2\,\text{m}^2, \ w = 9 \cdot 81 \times 10^3\,\text{N/m}^3$$

$$\text{Power of jet} = \frac{9 \cdot 81 \times 10^3 \times \tfrac{1}{4}\pi(0 \cdot 075)^2 \times 22 \cdot 5^3}{2g} \ W = \mathbf{25 \cdot 2\,kW}$$

$$\text{Velocity head at nozzle} = \frac{v^2}{2g} = \frac{22 \cdot 5^2}{2g} = 25 \cdot 9 \, \text{m}$$

$$\text{Loss of head in pipe line and nozzle} = 30 - 25 \cdot 9 = \mathbf{4 \cdot 1 \, m}$$

$$\text{Potential energy per unit wt of water in reservoir} = H$$

where H = height of reservoir surface above nozzle.

$$\text{Power supplied from reservoir} = WH$$

$$\text{Power delivered by jet} = W(v^2/2g)$$

$$\text{Efficiency of transmission} = \frac{W(v^2/2g)}{WH}$$

$$= \frac{v^2}{2gH} = \frac{22 \cdot 5^2}{2g \times 30}$$

$$= \mathbf{86 \cdot 3 \, per \, cent}$$

Problems

1 If $54 \cdot 5 \, \text{dm}^3$ of water are discharged from a vessel in 25 sec, find the rate of discharge in m^3/s.

If the discharge took place through an opening 50 mm diam, find the velocity of discharge.

Answer $0 \cdot 00218 \, \text{m}^3/\text{s}, \, 1 \cdot 11 \, \text{m/s}$

2 A piston of 46 mm diam slides concentrically in a fixed cylinder of 50 mm diam. The cylinder is filled with water and when the piston moves into the cylinder the water flows through the annular gap surrounding the piston. If the velocity of the piston is 75 mm/s relative to the cylinder, what is the velocity of flow through the gap.

Answer $0 \cdot 49 \, \text{m/s}$

3 Air enters a compressor with a density of $1 \cdot 2 \, \text{kg/m}^3$ at a mean velocity of 4 m/s in the 4 cm $\times$ 4 cm square inlet duct. Air is discharged from the compressor with a mean velocity of 3 m/s in a $2 \cdot 5$ cm diameter circular pipe. Determine the density at outlet and the mass flow rate.

Answer $5 \cdot 22 \, \text{kg/m}^3, \, 7 \cdot 68 \times 10^{-3} \, \text{kg/s}$

4 A 20 mm diam pipe forks, one branch being 10 mm in diam and the other 15 mm in diam. If the velocity in the 10 mm pipe is $0 \cdot 3$ m/s and that in the 15 mm pipe is $0 \cdot 6$ m/s, calculate the velocity in m/s and the rate of flow in cm^3/s in the 20 mm diam pipe.

Answer $129 \cdot 6 \, \text{cm}^3/\text{s}, \, 0 \cdot 413 \, \text{m/s}$

5 A fluid of constant density flows at the rate of 15 litres/sec along a pipe AB of 100 mm diam. This pipe branches at B into two pipes BC and BD each of 25 mm diam and a third pipe BE of 50 mm diam. The flow rates are such that the flow through BC is

three times the flow rate through BE and the velocity through BD is 4 m/s. Find the flow rates in the three branches BC, BD and BE and the velocities in pipes AB, BC and BE.
Answer 9·77 dm³/s, 1·96 dm³/s, 3·26 dm³/s, 1·91 m/s, 19·92 m/s, 1·66 m/s

6 Define the following terms: (*a*) steady flow, (*b*) unsteady flow, (*c*) streamline, (*d*) stream tube.

An incompressible fluid flows through a converging duct of circular cross-section. The diameter of the duct changes linearly over a length of 3·0 m, being 0·46 m at the entry and 0·15 m at the outlet. If the flow is steady and the volume rate of flow is 0·3 m³/s, determine the rate of acceleration of the fluid at a point halfway along the duct.

Determine the total rate of acceleration at the same point if the volume rate of flow increases at 0·37 m³/s².
Answer 11·32 m/s², 16·38 m/s²

7 Water is flowing along a pipe with a velocity of 7·2 m/s. Express this as a velocity head in metres of water. What is the corresponding pressure in kN/m².
Answer 2·64 m, 25·85 kN/m²

8 Water in a pipeline 36 m above sea level is under a pressure of 410 kN/m² and the velocity of flow is 4·8 m/s. Calculate the total energy per unit weight reckoned above sea level.
Answer 78·96 J/N

9 The suction pipe of a pump rises at a slope of 1 vertical in 5 along the pipe and water passes through it at 1·8 m/s. If dissolved air is released when pressure falls to more than 70 kN/m² below atmospheric pressure, find the greatest practicable length of pipe, neglecting friction. Assume that the water in the sump is at rest.
Answer 34·9 m

10 A helicopter of all-up weight 2 metric tons has a rotor 12 m in diameter. Find the vertical velocity of air passing through the rotor disc when the helicopter is hovering.

If the resistance to forward motion is $1·48V^2$N, find the inclination of the rotor to the horizontal to maintain a level cruising speed V of 18 m/s. Density of air is 1·28 kg/m³.
Answer 11·9 m/s, 88° 36′

11 A pipe of 150 mm bore is delivering water at the rate of 7500 dm³/min at a pressure of 820 kN/m². It connects by a gradually expanding pipe to a main of 300 mm bore which runs 3 m above it. Find the pressure in the 300 mm main, neglecting losses due to friction.
Answer 814 kN/m²

12 Water flows from a reservoir into a closed tank in which the pressure is 70 kN/m² below atmospheric. If the water level in the reservoir is 6 m above that in the tank, find the velocity of the water entering the tank, neglecting friction.
Answer 16·05 m/s

13 A pipe 300 m long tapers from 1·2 m diam to 0·6 m diam at its lower end and slopes downward at 1 in 100. The pressure at the upper end is 69 kN/m². Neglecting friction losses, find the pressure at the lower end when the rate of flow is 5·5 m³/min.
Answer 98·5 kN/m²

14 The diameter of a pipe changes gradually from 150 mm at a point A, 6 m above datum, to 75 mm at B, 3 m above datum. The pressure at A is 103 kN/m² and the velocity of flow is 3·6 m/s. Neglecting losses, determine the pressure at B.
Answer 35·3 kN/m²

15 A pipe, whose axis is horizontal, is full of water in motion. At a section A the velocity of the water is 90 m/min and the pressure is 138 kN/m. If the pipe tapers gradually from 150 mm diam at A to 100 mm diam at B, determine the pressure of the water at B, assuming that there is no loss of energy.
What must be the diameter of the pipe at B if the pressure there is reduced to 27·6 kN/m²?
Answer 133 kN/m², 47·6 mm

16 A jet of water is directed vertically up from a point A at which its diameter 75 mm and its velocity 9 m/s. Neglecting friction and assuming that the jet remains circular in section, what will be its diameter at a point 3 m above A?
Answer 102·6 mm

17 A jet of 75 mm diam at its base rises vertically 18 m. Find its diameter at a height of 12 m.
Answer 98·7 mm

18 Derive Bernoulli's equation from consideration of the forces and momentum changes as a fluid flows through a stream tube.
A jet of water 75 mm diameter at its base rises vertically to a height of 18 m. Obtain an equation for the diameter of the jet at any height above the base assuming that the jet does not break up and neglecting all energy losses. What is the diameter of the jet 10 m above the base.
Answer 91·85 mm

19 A vertical pipe carrying liquid of sp. gr. 0·83 upwards from a pump contracts gradually from 105 mm diam to 35 mm diam over a length of 0·45 m. A water U-tube is connected to two points 0·6 m apart in the two sections of the pipe and the head difference registered is 0·43 m. The connecting pipes to the U-tube are filled with the liquid that is flowing. Find the rate of flow in dm³/min. Ignore losses of energy.
Answer 76·4 dm³/min

20 Water flows out of a tank through a siphon formed by a bent pipe ABC, 25 mm in diam. The water in the tank is 1·5 m deep and the end of the pipe A is 0·3 m above the bottom of the tank. The length of pipe AB is vertical and is 8·7 m long, while BC is 17·4 m long. The pipe discharges into the atmosphere at C, 3 m below the bottom of the tank.

Assuming that the barometric pressure is equivalent to $10 \cdot 2$ m of water and that the loss of head in friction is $40v^2/2g$, where v is the velocity of flow in the pipe, calculate the rate of discharge and absolute pressure at B.

Answer $0 \cdot 72$ dm³/s, $1 \cdot 13$ m of water

21 Water is siphoned out of a tank by means of a bent pipe ABC 24 m long and 25 mm diam. The end A is below the water surface and 150 mm above the base of the tank. The length AB is vertical and 9 m long and BC is 15 m long with the discharge end C $1 \cdot 5$ m below the base of the tank. Assuming a barometric pressure of $10 \cdot 3$ m of water and that siphon action at B ceases when the absolute pressure is $1 \cdot 8$ m of water, determine the limiting velocity of water in the pipe and the depth of water in the tank when siphon action ceases. Assume that the loss of head in friction per metre is $0 \cdot 5v^2/2g$ where v = velocity in the pipe.

Answer $2 \cdot 37$ m/s, $2 \cdot 08$ m

22 Two sections of a pipe have areas of $0 \cdot 184$ m² and $0 \cdot 138$ m² respectively and the centre of the first is $3 \cdot 6$ m above that of the second. The pressure head at each section is $4 \cdot 5$ m and $0 \cdot 51$ m³/s of water flow from the higher to the lower section. Find the energy lost per kilogramme of flow.

Answer $32 \cdot 25$ J/kg

23 A vertical pipe conveying water tapers from 50 mm in diam at the top to 25 mm in diam at the bottom in a length of $1 \cdot 8$ m. A pressure gauge is connected at the top section and a second gauge is fitted at the bottom section. When $0 \cdot 194$ m³ of water per min flow up the pipe the gauges show a pressure difference of 31 kN/m². Determine the quantity of water which must flow downwards through the pipe if the gauges are to show no pressure difference and the frictional losses are assumed to vary as the square of the velocity.

Answer $111 \cdot 5$ dm³/min

24 In a vertical pipe conveying water, pressure gauges are inserted at A and B where the diameters are 150 mm and 75 mm respectively. The point B is $2 \cdot 4$ m below A and when the rate of flow down the pipe is 21 dm³/s the pressure at B is 12 kN/m² greater than at A. Assuming that the losses in the pipe between A and B can be expressed as $kv^2/2g$ where v is the velocity at A, find the value of k.

If the gauges at A and B are replaced by tubes filled with water and connected to a U-tube containing mercury of sp. gr. $13 \cdot 6$, give a sketch showing how the levels in the two limbs differ and calculate the value of this difference measured in millimetres.

Answer $35 \cdot 3$, $93 \cdot 6$ mm

25 The pressure in an hydraulic accumulator is 1250 kN/m². If 225 dm³ of water are pumped into the cylinder in 5 min, what is the work done and the power required?

Answer 281 000 J, 936 W

26 A pump draws water through a 150 mm diam pipe from a reservoir whose surface level is at datum level and discharges it through a 100 mm pipe to another reservoir whose surface level is 72 m above datum level. The pump is situated 6 m below datum. The loss of head in the 150 mm suction pipe is 3 times the velocity head in the 150 mm pipe and the loss of head in the 100 mm delivery pipe is 20 times the velocity head in the 100 mm pipe. Calculate the power output of the pump and the pressure heads at the inlet and outlet of the pump when the discharge is (*a*) $0 \cdot 91$ m³/min, (*b*) $2 \cdot 73$ m³/min.

Answers (*a*) $11 \cdot 28$ kW, $5 \cdot 85$ m, $81 \cdot 61$ m, (*b*) $47 \cdot 4$ kW, $4 \cdot 65$ m, $110 \cdot 5$ m

6

Flow measurement – venturi meter and pitot tube

Both the venturi meter and the pitot tube are examples of the practical application of Bernoulli's equation to the measurement of flow. In any practical hydraulic system loss of energy will occur, but it is convenient to ignore this loss in deriving the equations for these instruments and then to correct the theoretical result obtained by multiplying it by an experimentally determined coefficient to allow for the effect of loss of energy.

Venturi meter

6.1 Horizontal venturi meter

> (*a*) Describe the arrangement of a venturi meter and explain its mode of action.
>
> (*b*) Derive an expression for the theoretical discharge through a horizontal venturi meter and show how it must be modified to obtain the actual discharge.
>
> (*c*) A venturi tube tapers from 300 mm in diam at the entrance to 100 mm in diam at the throat, and the discharge coefficient is 0.98. A differential mercury U-tube gauge is connected between pressure tappings at the entrance and the throat. If the meter is used to measure the flow of water and the water fills the leads to the U-tube and is in contact with the mercury, calculate the discharge when the difference of level in the U-tube is 55 mm.

Solution. (a) The venturi meter consists of a short converging conical tube (Fig. 6.1) leading to a cylindrical portion called the "throat" which is followed by a diverging section. The entrance and exit diameter is the same as that of the pipe line into which it is inserted. The angle of the convergent cone is usually 21°, the length of throat is equal to the throat diameter, and the angle of the divergent cone is 5° to 7° to ensure a minimum loss of energy, but where this is unimportant this angle may be as large as 14°. Pressure tappings are taken at the entrance and at the throat, either from single holes or by using a number of holes around the circumference connecting to an annular chamber or piezometer ring, and the pressure difference is measured by a suitable gauge.

For continuity of flow the velocity v_1 at the entry section 1 will be less than the velocity v_2 at the throat section 2 since $a_1v_1 = a_2v_2$ and a_1 is greater than a_2. The kinetic energy in the throat will be *greater* than at the entrance and since by Bernoulli's theorem the total energy at the two sections is the same, the pressure energy at the throat will be less than at the entrance. The pressure difference thus created is dependent on the rate of flow through the meter.

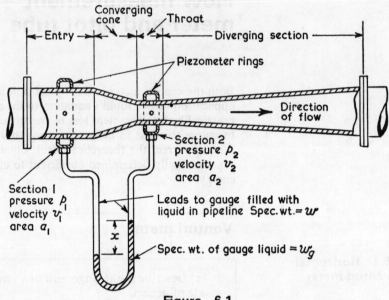

Figure 6.1

(*b*) Bernoulli's equation for sections 1 and 2 gives:

$$z_1 + \frac{v_1^2}{2g} + \frac{p_1}{w} = z_2 + \frac{v_2^2}{2g} + \frac{p_2}{w}$$

ignoring losses. For a horizontal meter $z_1 = z_2$

$$\frac{v_2^2 - v_1^2}{2g} = \frac{p_1 - p_2}{w} \qquad . \qquad . \qquad . \qquad . \quad (1)$$

For continuity of flow, $a_1v_1 = a_2v_2$, giving

$$v_2 = \frac{a_1}{a_2} v_1$$

Substituting in equation (1)

$$v_1^2 \left(\frac{a_1^2}{a_2^2} - 1 \right) = 2g \left(\frac{p_1 - p_2}{w} \right)$$

$$v_1 = \frac{a_2}{\sqrt{(a_1^2 - a_2^2)}} \sqrt{\left[2g \left(\frac{p_1 - p_2}{w} \right) \right]}$$

Discharge $Q = a_1v_1 = \dfrac{a_1a_2}{\sqrt{(a_1^2 - a_2^2)}} \sqrt{(2gH)} \qquad . \qquad . \qquad . \quad (2)$

where $H = (p_1 - p_2)/w$ = pressure difference expressed as a head of the liquid flowing in the meter.

If the area ratio $a_1/a_2 = m$, equation (2) becomes

$$Q = a_1 \sqrt{\frac{2gH}{m^2 - 1}}$$

The theoretical discharge Q can be converted to actual discharge by multiplying by the coefficient of discharge C_d found experimentally.

$$\text{Actual discharge} = C_d \times Q = C_d a_1 \sqrt{\frac{2gH}{m^2 - 1}} \qquad . \quad (3)$$

(c) If the leads of the U-tube are filled with water,

$$p_1 - p_2 = x(w_g - w)$$

$$\therefore \qquad H = \frac{p_1 - p_2}{w} = x\left(\frac{w_g}{w} - 1\right)$$

$x = 55\text{mm} = 0\cdot055\,\text{m}$ and for water and mercury $w_g/w = 13\cdot6$.

$$H = 0\cdot055 \times 12\cdot6 = 0\cdot693\,\text{m of water}$$

$$C_d = 0\cdot98,\ a_1 = \tfrac{1}{4}\pi d_1{}^2 = \tfrac{1}{4}\pi(0\cdot3)^2 = 0\cdot0706\,\text{m}^2,$$

$$m = \frac{a_1}{a_2} = \frac{d_1{}^2}{d_2{}^2} = \left(\frac{12}{4}\right)^2 = 9$$

Using equation (3),

$$\text{Actual discharge} = 0\cdot98 \times 0\cdot0706 \sqrt{\frac{2 \times 9\cdot81 \times 0\cdot693}{81 - 1}}$$

$$= 0\cdot0285\,\text{m}^3/\text{s}$$

6.2 Horizontal meter with oil flowing

A horizontal venturi meter measures the flow of oil of specific gravity $0\cdot9$ in a 75 mm diam pipe line. If the difference of pressure between the full bore and the throat tappings is $34\cdot5$ kN/m² and the area ratio m is 4, calculate the rate of flow, assuming a coefficient of discharge of $0\cdot97$.

Solution. From equation (3) of Example 6.1,

$$Q = C_d a_1 \sqrt{\frac{2gH}{m^2 - 1}}$$

The difference of pressure head H must be expressed in terms of the liquid flowing through the meter

$$H = \frac{p}{w} = \frac{34\cdot5 \times 10^3}{0\cdot9 \times 9\cdot81 \times 10^3} = 3\cdot92\,\text{m of oil}$$

$$a_1 = \tfrac{1}{4}\pi d^2 = \tfrac{1}{4}\pi(0\cdot075)^2 = 0\cdot00441\,\text{m}^2,\ m = 4,\ C_d = 0\cdot97$$

$$Q = 0\cdot97 \times 0\cdot00441 \sqrt{\frac{2 \times 9\cdot81 \times 3\cdot92}{16 - 1}} = 0\cdot0106\,\text{m}^3/\text{s}$$

6.3 Inclined venturi meter

Derive an expression for the rate of flow through an inclined venturi meter and show that, if a U-tube type of gauge is used to measure the pressure difference, the gauge reading will be the same for a given discharge irrespective of the inclination of the meter.

A vertical venturi meter measures the flow of oil of specific gravity 0·82 and has an entrance of 125 mm diam and a throat of 50 mm diam. There are pressure gauges at the entrance and at the throat, which is 300 mm above the entrance. If the coefficient for the meter if 0·97 and the flow in m³/s when the pressure difference is 27·5 kN/m².

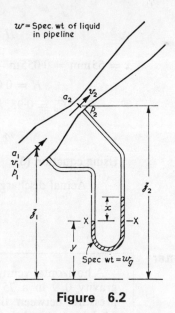

Figure 6.2

Solution. In Fig. 6.2 at the entrance to the meter the area, velocity pressure and elevation are a_1, v_1, p_1 and z_1 and at the throat the corresponding values are a_2, v_2, p_2 and z_2. From Bernoulli's equation,

$$z_1 + \frac{p_1}{w} + \frac{v_1^2}{2g} = z_2 + \frac{p_2}{w} + \frac{v_2^2}{2g}$$

$$v_2^2 - v_1^2 = 2g\left\{\left(\frac{p_1 - p_2}{w}\right) + (z_1 - z_2)\right\} \qquad . \qquad . \quad (1)$$

For continuity of flow $a_1v_1 = a_2v_2$ or $v_2 = mv_1$ where $m = $ area ratio $= a_1/a_2$. Substituting in equation (1) and solving for v_1

$$v_1 = \frac{1}{\sqrt{(m^2 - 1)}} \sqrt{\left[2g\left\{\left(\frac{p_1 - p_2}{w}\right) + (z_1 - z_2)\right\}\right]}$$

Discharge $Q = C_d a_1 v_1$

$$= \frac{C_d a_1}{\sqrt{(m^2 - 1)}} \sqrt{\left[2g\left\{\left(\frac{p_1 - p_2}{w}\right) + (z_1 - z_2)\right\}\right]} \qquad . \quad (2)$$

where C_d = coefficient of discharge.

Considering the U-tube gauge and assuming that the connexions are filled with the liquid in the pipe line, pressures at level XX are the same in both limbs.

For the l.-h. limb, $\qquad p_z = p_1 + w(z_1 - y)$

For the r.-h. limb, $p_z = p_2 + w(z_2 - y - x) + w_g x$

Thus, $\qquad p_1 + wz_1 - wy = p_2 + wz_2 - wy - wx + w_g x$

$$\frac{p_1 - p_2}{w} + z_1 - z_2 = x\left(\frac{w_g}{w} - 1\right)$$

Equation (2) can therefore be written

$$Q = \frac{C_d a_1}{\sqrt{(m^2 - 1)}} \sqrt{\left[2gx\left(\frac{w_g}{w} - 1\right)\right]}$$

This is independent of z_1 and z_2, so that the gauge reading x for a given rate of flow Q does not depend on the inclination of the meter. In equation (2),

$$a_1 = \tfrac{1}{4}\pi(0\cdot125)^2 = 0\cdot01226\,\text{m}^2$$

$p_1 - p_2 = 27\cdot5 \times 10^3\,\text{kN/m}^2$, $w = 0\cdot82 \times 9\cdot81 \times 10^3\,\text{N/m}^3$, $z_1 - z_2$
$= -0\cdot3\,\text{m}$, $m = (125/50)^2 = 6\cdot25$, $C_d = 0\cdot97$

$$Q = \frac{0\cdot97 \times 0\cdot01226}{\sqrt{[(6\cdot25)^2 - 1]}} \sqrt{\left[2 \times 9\cdot81\left(\frac{27\cdot5 \times 10^3}{0\cdot82 \times 9\cdot81 \times 10^3} - 0\cdot3\right)\right]}$$

$$= 0\cdot01535\,\text{m}^3/\text{s}$$

6.4 Venturi contraction operating piston

The water supply to a gas water heater contracts from 10 mm in diam at A (Fig. 6.3) to 7 mm in diam at B. If the pipe is horizontal, calculate the difference in pressure between A and B when the velocity of the water at A is 4·5 m/s.

This pressure difference operates the gas control through connexions which are taken to a horizontal cylinder in which a piston of 20 mm diam moves. Ignoring friction and the area of the piston connecting rod, what is the force on the piston?

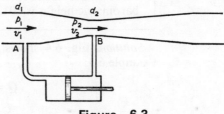

Figure   6.3

Solution. In Fig. 6.3 the diameter, pressure and velocity at A are d_1, p_1 and v_1, and at B they are d_2, p_2 and v_2.

By Bernoulli's theorem, for a horizontal pipe

$$\frac{p_1}{w} + \frac{v_1^2}{2g} = \frac{p_2}{w} + \frac{v_2^2}{2g}$$

$$\frac{p_1 - p_2}{w} = \frac{v_2^2 - v_1^2}{2g}$$

For continuity of flow $\frac{1}{4}\pi d_1^2 v_1 = \frac{1}{4}\pi d_2^2 v_2$

$$v_2 = v_1 \left(\frac{d_1}{d_2}\right)^2$$

Putting $v_1 = 4\cdot5\,\text{m/s}$, $d_1 = 10\,\text{mm}$, $d_2 = 7\,\text{mm}$

$$v_2 = 4\cdot5 \left(\frac{10}{7}\right)^2 = 9\cdot18\,\text{m/s}$$

and $\dfrac{p_1 - p_2}{w} = \dfrac{9\cdot18^2 - 4\cdot5^2}{2 \times 9\cdot81} = 3\cdot26\,\text{m of water}$

$$\text{Pressure difference} = p_1 - p_2 = 3\cdot26 \times 9\cdot81 \times 10^3\,\text{N/m}^2$$
$$= 31\cdot9\,\text{kN/m}^3$$

$$\text{Area of piston} = \tfrac{1}{4}\pi(0\cdot020)^2 = 0\cdot000314\,\text{m}^2$$

$$\text{Force on piston} = 31\cdot9 \times 10^3 \times 0\cdot000314 = \textbf{10·1N}$$

6.5 Venturi meter and inverted U-tube gauge

Water from a tank passes through a vertical venturi meter having a 125 mm diam inlet and 50 mm diam throat, the inlet being $0\cdot3$ m above the throat and 6 m below the level in the tank. The valve for controlling flow is on the outlet side of the meter. Tubes connect the inlet and throat to an inverted U-tube water gauge containing air in its upper portion. The zero on each limb of the gauge is $0\cdot6$ m below the venturi inlet and, when there is no flow, the gauge reading for each limb is $0\cdot9$ m, the volume of air contained in the upper part of the gauge being then equivalent to that contained in a 2 m length of the gauge tube. When the rate of flow is $12\cdot75$ dm^3, the gauge readings are $1\cdot71$ m and $-0\cdot57$ m. Determine (*a*) the discharge coefficient of the venturi, (*b*) the pressure head at the venturi throat, (*c*) the loss of head in the pipe between the tank and the venturi inlet. Assume that the volume of air in the gauge varies inversely as the pressure and that the barometric height is $10\cdot2$ m of water.

Solution. Fig. 6.4 shows the arrangement of the meter. (*a*) From Example 6.3:

$$Q = C_d a_1 \sqrt{\frac{2gh}{m^2 - 1}}$$

where m = area ratio and h = differential head as given by U-tube

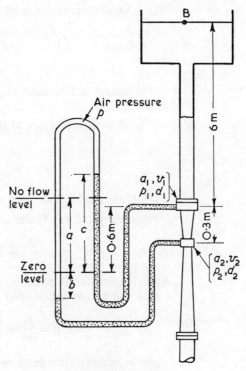

Figure 6.4

gauge. When $Q = 0.01275\,\text{m}^3/\text{s}$, $h = b + c = 1.71 + 0.57 = 2.28\,\text{m}$,
$a_1 = \frac{1}{4}\pi(0.125)^2 = 0.01225\,\text{m}^3$, $m = d_1{}^2/d_2{}^2 = (125/50)^2 = 6.25$
$m^2 = 39.1$

$$\therefore \qquad 0.01275 = C_d \times 0.01225 \sqrt{\frac{2 \times 9.81 \times 2.28}{38.1}}$$

$$C_d = \frac{0.01275}{0.01225 \times 1.085} = 0.96$$

(*b*) For no flow:

Volume of air in gauge $= 2A$

where A = area of tube.

Absolute pressure head of air in gauge

$$= \text{atmospheric head} + 6.6 - 0.9\,\text{m of water}$$
$$= 10.2 + 5.7 = 15.9\,\text{m of water}$$

For flow of $12.75\,\text{dm}^3/\text{s}$

Volume of air in gauge $= \{2 + (a + b) - (c - a)\}A$
$$= (2 + 2a + b - c)A$$
$$= (2 + 1.8 + 0.57 - 1.71)A = 2.66A$$

$\therefore$ Absolute pressure head of air in gauge $= H$

$$= \frac{2A}{2 \cdot 66A} \times 15 \cdot 9$$

$$= 11 \cdot 95 \, \text{m of water}$$

Since level in l.-h. limb of gauge is $b + 0 \cdot 3 = 0 \cdot 87 \, \text{m}$ below level of throat:

Absolute pressure at throat $= p_2 = w(H - 0 \cdot 87)$

$$= 9 \cdot 81 \times 10^3 (11 \cdot 95 - 0 \cdot 87)$$

$$= \mathbf{115 \cdot 8 \, kN/m^2}$$

(c) Since level in r.-h. limb of gauge is $c - 0 \cdot 6 = 1 \cdot 11 \, \text{m}$ above inlet

Absolute pressure at inlet $= p_1 = w(H + 1 \cdot 11) = 13 \cdot 06$

$\therefore$ $\frac{p_1}{w} = 13 \cdot 06 \, \text{m of water}$

Applying Bernoulli's equation to the free surface of the tank and to the throat and taking datum level as the inlet

$$\frac{p_B}{w} + 6 = \frac{p_1}{w} + \frac{v_1^2}{2g} + \text{loss of head in pipe}$$

$p_B/w = $ atmospheric head $= 10 \cdot 2 \, \text{m}$, $v_1 = Q/a_1 = 0 \cdot 01275/0 \cdot 01225$
$= 1 \cdot 04 \, \text{m/s}$

$$\text{loss of head in pipe} = (10 \cdot 2 + 6) - \left(13 \cdot 06 + \frac{1 \cdot 04^2}{2 \times 9 \cdot 81}\right)$$

$$= \mathbf{3 \cdot 08 \, m \, of \, water}$$

6.6 Friction loss in venturi meter

Show that in a venturi meter the quantity of water passing through the meter will only be proportional to the measured venturi head H if the head lost in friction h_f is proportional to the head difference h_v due to increased velocity.

A venturi meter has a coefficient of discharge of $0 \cdot 97$ and the frictional loss in the diverging cone is twice that in the converging cone. What will be the total head lost in friction in the meter when the measured difference of head is equivalent to 410 mm of water?

Solution. If $C_d = $ coefficient of discharge, the discharge will be

$$Q = C_d \frac{A\sqrt{(2g)}}{\sqrt{(m^2 - 1)}} H^{1/2} \qquad . \qquad . \qquad . \quad (1)$$

Since the coefficient of discharge is introduced to allow for the fact that the measured head H is greater than the true head due to velocity increase h_v, we also have

$$Q = \frac{A\sqrt{(2g)}}{\sqrt{(m^2 - 1)}} h_v^{1/2}$$

Therefore

$$C_d = \sqrt{\frac{h_v}{H}} \qquad . \qquad . \qquad . \qquad . \qquad . \qquad . \quad (2)$$

But

$$H = h_v + h_f$$

so that

$$C_d = \sqrt{\frac{h_v}{h_v + h_f}}$$

If Q is proportional to $H^{1/2}$ then the from equation (1) C_d must be constant and

$$\frac{h_v}{h_v + h_f} = K$$

where K is a constant, giving

$$h_f = \frac{1 - K}{K} h_v$$

or

$$h_f \propto h_v$$

From equation (2), $h_v = C_d^2 H$

Loss of head in converging cone $= h_f = H - h_v = H(1 - C_d^2)$

If loss of head in diverging cone $= 2h_f$

Total loss of head $= 3h_f = 3H(1 - C_d^2)$

Putting $H = 0{\cdot}41\,\text{m}$ and $C_d = 0{\cdot}97$

Total loss of head $= 3 \times 0{\cdot}41(1 - 0{\cdot}97^2)$

$\qquad\qquad\qquad\qquad = \mathbf{0{\cdot}0726\,m\ of\ water}$

Pitot tube

6.7   Pitot tube

Describe, with the help of neat diagrams, the construction and operation of (a) a pitot-static tube, and (b) a manometer suitable for use with such a tube when the difference of head is very small.

A pitot-static tube used to measure air velocity along a wind tunnel is coupled to a manometer which shows a difference of head of 4 mm water. The density of air is $1{\cdot}2\,\text{kg/m}^3$. Obtain the air velocity, assuming that the coefficient for the pitot tube is unity.

Solution. Fig. 6.5(a) shows a simple pitot tube. If the velocity of the stream at A is v a particle moving from A to the mouth of the tube B will be brought to rest so that v_0 at B is zero. By Bernoulli's theorem:

Total energy at A $=$ Total energy at B

$$\frac{v^2}{2g} + \frac{p}{w} = \frac{v_0^2}{2g} + \frac{p_0}{w} = \frac{p_0}{w} \qquad . \qquad . \qquad . \quad (1)$$

Now $p/w = d$ and the increased pressure at B will cause the liquid in the vertical limb of the pitot tube to rise to a height h above the free surface so that $p_0/w = h + d$.

From equation (1):

$$\frac{v^2}{2g} = \frac{p_0 - p}{w} = h \quad \text{or} \quad v = \sqrt{2gh}$$

Fig. 6.5(b) shows a pitot tube used in a pipe; $p_0 - p$ is the difference between the static pressure and the pressure at the impact hole and is measured by a differential gauge.

The static pressure tapping and the impact tapping can be combined in a single pitot-static tube. Fig. 6.5(c) shows an N.P.L. hemispherical head pitot-static tube. The inner tube measures the impact pressure and the outer sheath has 7 holes as shown, which are open to the static pressure. Leads are taken from the inner and outer tubes to a sensitive manometer. The impact hole should be accurately directed upstream and the instrument measures the velocity at the point at which the impact hole is placed in the cross-section.

Although theoretically $v = \sqrt{(2gh)}$, pitot tubes may require calibration. The actual velocity is then given by $v = C\sqrt{(2gh)}$ where C is the coefficient of the instrument. For the pitot-static tube shown in Fig. 6.5(c), the value of C is unity for values of Reynolds number exceeding 3000, taking d in the Reynolds number $\rho vd/\eta$ as the diameter of the pitot tip.

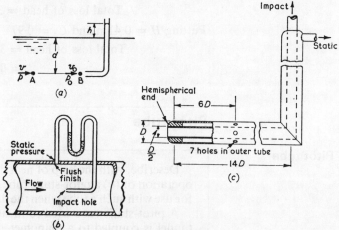

Figure 6.5

A suitable sensitive manometer for use with a pitot-static tube is shown in Fig. 6.6. It consists of a U-tube attached to a tilting beam which pivots on a knife edge under the left-hand limb and is tilted by adjusting a micrometer screw at the other end. The U-tube is filled with liquid as shown and with equal pressure on both limbs the level of the right-hand end of the beam is adjusted until the liquid surface is in line with the horizontal cross-wire of the microscope. The left-hand limb is now connected to the impact tapping and the right-hand limb to the static tapping of the pitot-static tube. The level of the liquid will rise in

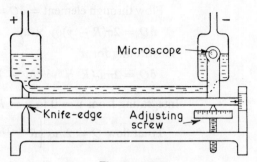

Figure 6.6

the right-hand limb which is now raised by the adjusting screw until the surface of the liquid once more coincides with the microscope cross-wire. The height through which the right-hand limb is raised is read off the micrometer scale and is the difference of head in terms of the liquid in the gauge.

$$\text{Velocity of air} = C\sqrt{(2gh)}$$

where C = coefficient of the tube = 1·0, h = difference of head expressed in terms of the fluid flowing, namely air = $0·004(\rho_w/\rho_a) = 0·004 \times (10^3/1·2) = 3·333$ m of air

$$\text{Velocity of air} = 1·0\sqrt{(2g \times 3·333)} = \mathbf{8·06\,m/s}$$

6.8 Position for pitot tube to measure mean velocity

The velocity distribution in a circular pipe of radius R can be expressed as

$$u = u_0 \left(\frac{y}{R}\right)^{1/m}$$

where u is the velocity at a point distant y from the wall and u_0 is the velocity at the axis.

Demonstrate that if a pitot tube is placed at $0·25R$ from the wall, the pitot registers the correct mean velocity within $\pm 0·5$ per cent for a range of m from 4 to 10.

Find the kinetic energy per unit weight of flow in terms of the mean velocity when $m = 7$.

Solution. To find the mean velocity V for any value of m, consider an annular element of width δy at a distance y from the wall (Fig. 6.7).

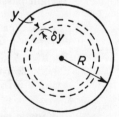

Figure 6.7

Flow through element $= \delta Q =$ area $\times$ velocity

$$\delta Q = 2\pi (R - y)\delta y \times u,$$

substituting for u

$$\delta Q = 2\pi u_0 (R^{-1/m} y^{1/m}\delta y - R^{-1/m} y^{(m+1)/m}\delta y)$$

Integrating from $y = 0$ to $y = R$:

$$\text{Total flow } Q = 2\pi u_0 \left\{ R^{(m-1)/m} \int_0^R y^{1/m} dy - R^{-1/m} \int_0^R y^{(m+1)/m} dy \right\}$$

$$= 2\pi u_0 R^2 \left(\frac{m}{m+1} - \frac{m}{2m+1} \right) = \pi R^2 \times \frac{2u_0 m^2}{(m+1)(2m+1)}$$

$$\text{Mean velocity } V = \frac{Q}{\pi R^2} = \frac{2u_0 m^2}{(m+.1)(2m+1)} \cdot \qquad . \qquad (1)$$

Velocity when $y = 0 \cdot 25R$ recorded by pitot tube is

$$U = u_0 \left(\frac{0 \cdot 25R}{R} \right)^{1/m} = u_0 \left(\frac{1}{4} \right)^{1/m}$$

When $m = 10$ $\qquad V = u_0 \times \frac{2 \times 10^2}{11 \times 21} = 0 \cdot 866 u_0$

$$U = u_0 (0 \cdot 25)^{1/10} = 0 \cdot 870$$

$$\text{Percentage error in } U = \frac{0 \cdot 870 - 0 \cdot 866}{0 \cdot 870} \times 100$$

$$= +0 \cdot 46 \text{ per cent}$$

When $m = 4$

$$V = u_0 \times \frac{2 \times 4^2}{5 \times 9} = 0 \cdot 710 u_0$$

$$U = u_0 (0 \cdot 25)^{1/4} = 0 \cdot 707 u_0$$

$$\text{Percentage error in } U = \frac{0 \cdot 707 - 0 \cdot 710}{0 \cdot 707} \times 100 = -0 \cdot 43 \text{ per cent}$$

In one second:

Weight passing per second through annular element $= w\delta Q$

$$\text{Kinetic energy per second of this fluid} = \left(\frac{1}{2} \right) \frac{w\delta Q}{g} u^2$$

For the whole cross-section

$$\text{Weight passing per second} = \int_{y=0}^{y=R} w\delta Q$$

$$= w \int_0^R 2\pi (R - y) u_0 \left(\frac{y}{R} \right)^{1/7} dy$$

$$= \frac{2\pi u_0 w}{R^{1/7}} \int_0^R (Ry^{1/7} - y^{8/7}) dy = 2\pi R^2 w \times 0 \cdot 409 u_0$$

$$\text{Kinetic energy per second of this fluid} = \int_{y=0}^{y=R} \frac{wu^2 \delta Q}{2g}$$

$$= \frac{w}{2g} \int_0^R 2\pi(R - y)u^3 dy$$

$$= \frac{2\pi w u_0^3}{2g R^{3/7}} \int_0^R (Ry^{3/7} - y^{10/7})dy$$

$$= 2\pi R^2 w \times 0.288 \frac{u_0^3}{2g}$$

$$\text{Kinetic energy per unit weight} = \frac{\text{Kinetic energy/sec}}{\text{Weight/sec}}$$

$$= \frac{0.288}{0.409} \frac{u_0^2}{2g}$$

Substituting for u_0 in terms of V from equation (1)

$$\text{Kinetic energy per unit weight} = \frac{0.288}{0.409} \left\{ \frac{(m + 1)(2m + 1)}{2m^2} \right\}^2 \frac{V^2}{2g}$$

and since $m = 7$

$$\text{Kinetic energy per unit weight} = \frac{0.288}{0.409} \left(\frac{8 \times 15}{98} \right)^2 \frac{V^2}{2g}$$

$$= 1.056 \frac{V^2}{2g}$$

Problems

1 A venturi meter measures the flow of water in a 75 mm diam pipe. The difference of head between the throat and the entrance of the meter is measured by a U-tube containing mercury, the mercury being in contact with the water. What should be the diameter of the throat of the meter in order that the difference in level of the mercury be 250 mm when the quantity of water flowing in the pipe is 620 dm³/min? Assume a coefficient of discharge of 0·97.
Answer 40·7 mm

2 A venturi meter is tested with its axis horizontal and the flow measured by means of a tank. The pipe diameter is 76 mm, the throat diameter is 38 mm and the pressure difference is measured by a U-tube containing mercury, the connexions being full of water. If the difference in levels in the U-tube remains steady at 266 mm of mercury while 2200 kg of water are collected in 4 min, what is the coefficient of discharge? (Sp. gr. of mercury is 13·6.)
Answer 0·966

3 A venturi meter fitted to a pipe of 450 mm bore has a throat diameter of 200 mm. Find the quantity of water flowing when the venturi head is 175 mm of water. Take $C_d = 0.96$.
Answer 0·057 m³/s

4 A venturi meter installed in a horizontal water main has a throat diameter of 75 mm and a pipe diameter of 150 mm. The coefficient of discharge is $0 \cdot 97$. Calculate the rate of flow in the main in m^3/h if the difference of level in a mercury U-tube gauge connected to the throat and full bore tappings is 178 mm, the mercury being in contact with the water.
Answer $106 \cdot 5$ m^3/h

5 What are the relative advantages of using a venturi meter to measure flow compared with an orifice meter?

A venturi meter has a main diameter of 65 mm and a throat diameter of 26 mm. When measuring the flow of a liquid of density 898 kg/m^3 the reading of a mercury differential-pressure gauge was 71 mm. Working from first principles or proving any formula used, calculate the flow through the meter in m^3/h. Take the coefficient of the meter as $0 \cdot 97$ and the sp. gr. of mercury as $13 \cdot 6$.
Answer $8 \cdot 36$ m^3/h

6 A venturi meter having an entrance diameter of 300 mm and a throat diameter of 200 mm is used to measure the volume of gas flowing through a pipe. The discharge coefficient is $0 \cdot 96$. Assuming the density of the gas to be constant and equal to $0 \cdot 99$ kg/m^3, calculate the volume flowing per sec when the pressure difference between entrance and throat as measured on a water U-tube gauge is 61 mm. Prove any formula used for the meter.
Answer $1 \cdot 17$ m^3/s

7 The throat and full bore diameters of a venturi meter are 19 mm and 57 mm respectively. Calculate the coefficient of discharge of the meter if the pressure at the full bore section is $172 \cdot 5$ kN/m^2 above that at the throat when the meter is passing 311 dm^3/min of water. The centreline of the meter is inclined to the horizontal, the throat section being $0 \cdot 46$ m above the full bore section.
Answer $0 \cdot 99$

8 A venturi meter has its axis vertical, the inlet and throat diameters being 150 mm and 75 mm respectively. The throat is 225 mm above the inlet and the coefficient of discharge is $0 \cdot 96$. Petrol of sp. gr. $0 \cdot 78$ flows through the meter at the rate of $39 \cdot 6$ dm^3/s. By application of Bernoulli's principle, find:
  (*a*) the pressure difference in kN/m^2 between inlet and throat;
  (*b*) the difference of level which would be registered by a vertical mercury manometer, the tubes above the mercury being full of petrol. (sp. gr of mercury = $13 \cdot 6$.)
Answer 34 kN/m^2, $0 \cdot 257$ m

9 A venturi meter in a horizontal 300 mm diam water pipe has a throat diameter of 100 mm and a discharge coefficient of $0 \cdot 98$. A mercury U-tube was used to measure the difference of head between the pipe inlet and the throat at points $0 \cdot 6$ m apart. If the mercury gauge reading was 760 mm, find the flow through the pipe in dm³/s. The connections to the U-tube are filled with water.

If the meter had been placed vertically with the flow upward, what would have been the gauge reading for the same rate of flow?
Answer 106·2 dm³/s, 760 mm

10 Oil of specific gravity 0·85 flows upwards through a vertical venturi meter fitted in a 225 mm diam pipeline. The diameter of the throat is 75 mm and the throat is 150 mm above the inlet.

Pressure connexions are taken from the inlet and throat of the meter to the lower and upper ends of a vertical cylinder in which a piston with an area of 1950 mm² is free to move. A load can be applied to the piston by placing weights on a piston rod passing through the bottom of the cylinder.

Calculate the discharge through the meter if a total load of 13·65 kg, including the piston, is required to maintain the piston in equilibrium. Neglect frictional losses in the converging section of the meter and also the cross-sectional area of the piston rod.
Answer 55·8 dm³/s

11 A venturi contraction is introduced into a 0·75 m diam horizontal pipe. The area of the pipe is 6 times that of the throat. The upper end of a vertical cylinder of 0·3 m diam is connected by a pipe to the throat, while the lower end is connected to the beginning of the convergence. Neglecting frictional losses and the thickness of the piston, determine the flow through the pipe in cubic metres per sec at which the piston begins to rise, when the effective load − piston, piston rod and external load − on the piston is 218 kg. The piston rod is 38 mm in diam and passes through both ends of the cylinder.
Answer 0·586 m³/s

12 A pitot-static tube placed in the centre of a 200 mm pipe line conveying water has one orifice pointing upstream and the other perpendicular to it. If the pressure difference between the two orifices is 38 mm of water when the discharge through the pipe is 22 dm³/s, calculate the meter coefficient. Take the mean velocity in the pipe to be 0·83 of the central velocity.
Answer 0·977

13 Obtain an expression for the velocity in terms of the observed difference in level of the liquid, of specific gravity S, in the U-tube connected to the up- and down-stream orifices of a pitot tube immersed in flowing water. If the difference of level is 0·36 m, the specific gravity of the liquid 1·25 and the calibration coefficient for the orifices 0·865, what is the velocity in m/s?
Answer 0·813 m/s

14 A pitot tube during calibration tests in water gave the following readings −

Velocity of fluid, v m/s	0·569	0·904	1·282	1·975	2·435
Head H cm of water	1·925	4·37	8·89	23·2	36·6

Find the value of the constant for the tube when v is in m/s and H is measured in centimetres.

Answer $(v = 0 \cdot 412 \sqrt{H}$

15 Show that if a pitot-static tube immersed in a stream of fluid of density ρ_1, is connected to a U-tube manometer containing a manometric fluid of density ρ_2, the velocity of the fluid flowing immediately upstream of the pitot-static tube is given by

$$v = k \left[2gh \left(\frac{\rho_2}{\rho_1} - 1 \right) \right]^{\frac{1}{2}} \text{ where } h \text{ is the manometric reading and } k$$

is a calibration constant.

A pitot-static tube is used to investigate the flow of air in a $0 \cdot 5$ m diameter circular duct. It is found that the velocity vanes over the cross section so that the velocity v at any radius r is

$$v = v_0 \left(1 - \frac{r^7}{a^7} \right) \text{ where } v_0 \text{ is the velocity at the duct centreline and}$$

a is the radius of the duct. If the reading on a water manometer connected to the pitot-static tube is 5 mm when the pitot-static tube is at the duct centreline, determine the volume flow rate. Asume k is unity.

Answer $1 \cdot 375$ m³/s

16 A pitot-static tube is located in the centre of an air duct to measure the velocity on the duct axis. An inclined manometer containing fluid of relative density $0 \cdot 785$ is connected to the pitot-static tube to measure the pressure difference between the tappings. If the inclined tube of the manometer makes an angle of 15 deg with the horizontal and the ratio of the diameters of the enlarged vertical tube to the inclined tube is 5, find the decrease in velocity in the duct if the manometer deflection along the inclined tube decreases from 150 mm to 100 mm.

Answer $4 \cdot 4$ m/s

7

Flow measurement – small and large orifices

An orifice is an opening in the side or base of a tank or reservoir. It is completely below the surface, usually circular and the rate of flow or discharge depends upon the head of liquid above the orifice.

An orifice is called a small orifice when it has a diameter that is small compared with the head producing flow so that the velocity through the orifice does not vary appreciably from top to bottom of the opening.

7.1 Torricelli's theorem

Derive from Bernoulli's theorem expressions for the theoretical velocity and discharge through a small orifice. How does the actual discharge compare with this theoretical value? Explain the reasons for the difference.

A sharp-edged orifice of 50 mm diam discharges water under a head of $4 \cdot 5$ m. Find the coefficient of discharge if the measured rate of flow is $11 \cdot 45$ dm³/s. If there is an average pressure within the jet in the plane of the orifice of $26 \cdot 5$ kN/m² above atmospheric pressure calculate the coefficient of contraction. Neglect loss of energy due to friction.

Solution. Fig. 7.1(*a*) shows a small orifice in the side of a large reservoir. At a point A on the free surface the pressure p_A is atmospheric and, if the tank is large, the velocity v_A will be negligible. At a point B just outside the orifice the velocity of the jet $v_B = v$; the pressure p_B is atmospheric so that $p_B = p_A$. Taking the datum level for potential energy at the centre of the orifice and applying Bernoulli's equation to A and B.

$$z_A + \frac{v_A{}^2}{2g} + \frac{p_A}{w} = z_B + \frac{v_B{}^2}{2g} + \frac{p_B}{w}$$

Putting $z_A - z_B = h$, $v_A = 0$, $v_B = v$ and $p_A = p_B$,

$$\text{Velocity of jet} = v = \sqrt{(2gh)} \qquad . \qquad . \qquad . \quad (1)$$

This is a statement of Torricelli's theorem.

Theoretical discharge through orifice = area × velocity

If A = area of jet at C, then

$$Q = A\sqrt{(2gh)} \qquad . \qquad . \qquad . \quad (2)$$

In practice the actual discharge is considerably less than the theoretical value given in equation (2) for two reasons. The velocity of the jet is less than that given by equation (1) because of frictional resistance.

$$\text{Actual velocity } v' = C_v \times v = C_v\sqrt{(2gh)}$$

where C_v = coefficient of velocity.

Also as shown in Fig. 7.1(b) the paths of the particles of the liquid converge on the orifice so that the area of the issuing jet is less than the

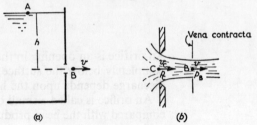

Figure 7.1

area of the orifice. In the plane of the orifice the particles have a component of velocity towards the centre so that at C the pressure is greater than atmospheric pressure. It is only at B a little outside the orifice that the paths of the particles become parallel. The section through B is called the *vena contracta*.

$$\text{Actual area of jet } A' = C_c \times A$$

where C_c = coefficient of contraction.

$$\text{Actual discharge = actual area} \times \text{actual velocity}$$
$$= C_c A \times C_v\sqrt{(2gh)}$$

or putting $C_c \times C_v = C_d$ = coefficient of discharge

$$\text{Actual discharge} = C_d A\sqrt{(2gh)} \quad . \qquad . \qquad . \quad (3)$$

From equation (3) $\qquad C_d = \dfrac{Q}{A\sqrt{(2gh)}}$

Putting $Q = 11 \cdot 45 \text{dm}^3/\text{s}$, $A = \frac{1}{4}\pi(0 \cdot 05)^2 = 19 \cdot 6 \times 10^{-4}\text{m}^2$, $h = 4 \cdot 5\text{m}$

$$C_d = \frac{11 \cdot 45 \times 10^{-3}}{1 \cdot 96 \times 10^{-3}\sqrt{(2 \times 9 \cdot 81 \times 4 \cdot 5)}} = \mathbf{0 \cdot 62}$$

Also applying Bernoulli's equation to B and C in Fig. 7.1(b):

$$\frac{v^2}{2g} + \frac{p_B}{w} = \frac{v_C^2}{2g} + \frac{p_C}{w} \quad . \qquad . \qquad . \quad (4)$$

From equation (1), ignoring friction,

$$v = \sqrt{(2gh)}$$
$$\therefore \qquad v = \sqrt{(2g \times 4 \cdot 5)} = 9 \cdot 38\,\text{m/s}$$
$$\text{Also} \qquad p_C - p_B = 26 \cdot 5 \times 10^3\,\text{N/m}^2$$
$$w = \rho g = 10^3 \times 9 \cdot 81\,\text{N/m}^3$$

From equation (4)

$$v_c{}^2 = v^2 - 2g\left(\frac{p_C - p_B}{w}\right) = 9\cdot38^2 - \frac{2 \times 9\cdot81 \times 26\cdot5 \times 10^3}{10^3 \times 9\cdot81}$$

$$= 88\cdot3 - 53\cdot0 = 35\cdot3$$

$$v_c = 5\cdot94\text{m/s}$$

For continuity of flow $A_B v = A_C v_c$ where A_B and A_C are the areas of the jet at B and C.

$$\text{Coefficient of contraction} = \frac{A_B}{A_C} = \frac{v_c}{v} = \frac{5\cdot94}{9\cdot38} = \mathbf{0\cdot634}$$

7.2 Coefficients and types of orifice

Define the coefficients used in connexion with flow through orifices, explaining why these coefficients are necessary.

Describe, with sketches, three different types of orifices, indicating the approximate values of the coefficient of discharge for each.

A 25 mm diam nozzle discharges $0\cdot76$ m³ of water per minute when the head is 60 m. The diam of the jet is $22\cdot5$ mm. Determine (a) the value of the coefficients, (b) the loss of head due to fluid resistance.

Solution.

$$\text{Coefficient of velocity } C_v = \frac{\text{Actual velocity at } vena\ contracta}{\text{Theoretical velocity}}$$

or

$$\text{Actual velocity} = C_v \sqrt{(2gh)}$$

$$\text{Coefficient of contraction } C_c = \frac{\text{Area of jet at } vena\ contracta}{\text{Area of orifice}}$$

$$\text{Coefficient of discharge } C_d = \frac{\text{Actual discharge}}{\text{Theoretical discharge}}$$

The explanation of the need for these coefficients is given in Example 7.1.

Fig. 7.2 shows a sharp-edged orifice, a rounded orifice and a Borda re-entrant orifice. The latter has two types of flow (a) running free with the jet clear of the sides of the tube, and (b) running full with the jet touching the sides of the tube at the outlet. Coefficients of discharge are indicated in the diagram.

(a) Actual discharge $= 0\cdot76$ m³/min $= 0\cdot0127$ m³/s. If A = area of orifice and h = head producing flow,

$$\text{Theoretical discharge} = A\sqrt{(2gh)} = \tfrac{1}{4}\pi(0\cdot025)^2\sqrt{(2g \times 60)}$$

$$= 4\cdot9 \times 10^{-4} \times 34\cdot2 = 0\cdot0168\,\text{m}^3/\text{s}$$

$$\text{Coefficient of discharge} = \frac{0\cdot0127}{0\cdot0168} = \mathbf{0\cdot758}$$

$$\text{Area of nozzle} = \tfrac{1}{4}\pi \times (0{\cdot}025)^2 \text{m}^2$$

$$\text{Area of jet} = \tfrac{1}{4}\pi \times (0{\cdot}0225)^2 \text{m}^2$$

$$\text{Coefficient of contraction} = \frac{\tfrac{1}{4}\pi \times (0{\cdot}0225)^2}{\tfrac{1}{4}\pi \times (0{\cdot}025)^2} = \mathbf{0{\cdot}810}$$

$$\text{Coefficient of velocity} = \frac{\text{Coefficient of contraction}}{\text{Coefficient of discharge}}$$

$$= \frac{0{\cdot}758}{0{\cdot}810} = \mathbf{0{\cdot}937}$$

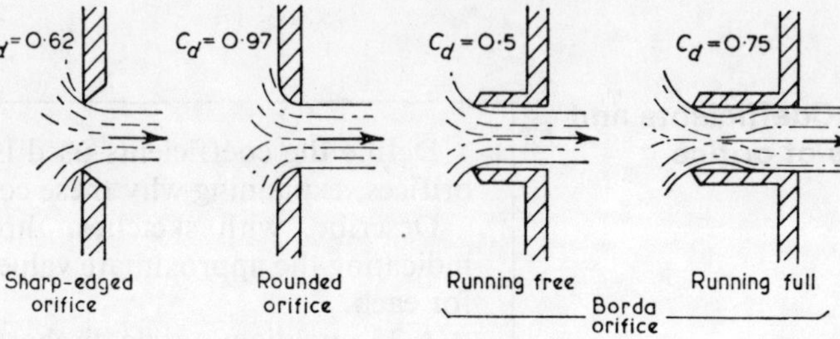

$C_d = 0{\cdot}62$ $C_d = 0{\cdot}97$ $C_d = 0{\cdot}5$ $C_d = 0{\cdot}75$

Sharp-edged orifice   Rounded orifice   Running free   Running full

Borda orifice

Figure 7.2

(b) Potential energy at nozzle per unit weight $= h$

$$\text{Kinetic energy of jet} = \frac{v^2}{2g}$$

where $v =$ jet velocity.

$$\text{Loss of energy per unit weight} = h - \frac{v^2}{2g}$$

But $$v = C_v\sqrt{(2gh)}$$

$$\text{Loss of energy per unit weight} = h(1 - C_v{}^2)$$

This is the head lost due to fluid resistance.

$$\text{Loss of head} = 60(1 - 0{\cdot}937^2) = \mathbf{7{\cdot}32m \ of \ water}$$

7.3 Profile of jet

A jet of liquid issues horizontally from an orifice in the vertical side of the tank. Derive an expression for the actual velocity of the jet v if the jet falls a distance y vertically in a horizontal distance x measured from the *vena contracta*. If the head of liquid above the orifice is h, what will be the coefficient of velocity?

A circular orifice, 650 mm² in area, is made in the vertical side of a tank. If the jet falls vertically through $0{\cdot}5$ m while moving horizontally through $1{\cdot}5$ m and the discharge is $0{\cdot}10$ m³/min of water, determine the horizontal reaction of the jet on the tank.

Solution. The jet issues as shown in Fig. 7.3. If $t =$ time for a particle to travel from the *vena contracta* A to the point B, $x = vt$ and $y = \tfrac{1}{2}gt^2$

$$v = \frac{x}{t}$$

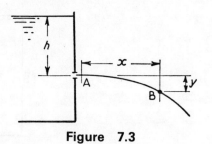

Figure 7.3

but
$$t = \sqrt{\frac{2y}{g}}$$

$$v = \sqrt{\frac{gx^2}{2y}} \quad . \quad . \quad . \quad . \quad . \quad (1)$$

From Example 7.1; theoretical velocity $= \sqrt{(2gh)}$

$$\text{Coefficient of velocity} = \frac{\text{Actual velocity}}{\sqrt{(2gh)}}$$

$$= \frac{v}{\sqrt{(2gh)}} = \sqrt{\frac{x^2}{4yh}}$$

Putting $x = 1\cdot5$m and $y = 0\cdot5$m in equation (1)

$$v = \sqrt{\frac{9\cdot81 \times 2\cdot25}{2 \times 0\cdot5}} = 4\cdot7\text{m/s}$$

As the particles of liquid leave the orifice they acquire momentum and therefore exert a reaction equal to their rate of change of momentum.

Horizontal reaction = rate of change of momentum

= mass/sec × velocity received

$$\text{Mass discharged/sec} = \rho Q = 10^3 \times \frac{0\cdot10}{60} = 1\cdot67\text{kg/s}$$

Reaction of jet $= 1\cdot67 \times 4\cdot7 = \mathbf{7\cdot85N}$

7.4 Maximum throw of jet

A tank $1\cdot8$ m high, standing on the ground, is kept full of water. There is an orifice in its vertical side at a depth h m below the surface. Find the value of h in order that the jet may strike the ground at a maximum distance from the tank.

Solution. If v is the actual velocity of the issuing jet, then, referring to Fig. 7.4, $v = C_v\sqrt{(2gh)}$.

Let t be the time taken for a particle to travel from the orifice to the ground, then

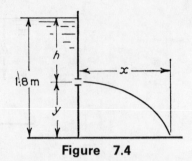

Figure 7.4

Horizontal distance travelled $= x = vt$

Vertical distance travelled $= y = \frac{1}{2}gt^2$

Eliminating t these equations give

$$x = \sqrt{\frac{2v^2y}{g}}$$

Putting $y = 1\cdot8 - h$ and $v = C_v\sqrt{(2gh)}$

$$x = \sqrt{\frac{C_v^2 4gh(1\cdot8 - h)}{g}} = 2C_v\sqrt{[h(1\cdot8 - h)]}$$

Thus x will be a maximum when $h(1\cdot8 - h)$ is a maximum or

$$\frac{d[h(1\cdot8 - h)]}{dh} = 1\cdot8 - 2h = 0$$

$$h = 0\cdot9\text{m}$$

7.5 Free surface under pressure

The water in a tank is $1\cdot8$ m deep and over the surface is air at pressure 70 kN/m² above atmosphere. Find the rate of flow in m³/s from an orifice of 50 mm diam in the bottom of the tank, given that $C_d = 0\cdot6$.

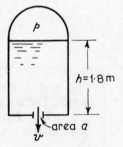

Figure 7.5

Solution. Referring to Fig. 7.5, apply Bernoulli's theorem to the surface of the water and the outlet.

$$\frac{p}{w} + h = \frac{v^2}{2g}$$

where p is the pressure of air in the tank above atmospheric pressure.

$$\text{Theoretical velocity } v = \sqrt{2g\left(\frac{p}{w} + h\right)}$$

$$\text{Actual rate of flow} = Q = C_d av = C_d a\sqrt{2g\left(\frac{p}{w} + h\right)}$$

Putting $C_d = 0{\cdot}6$, $a = \frac{1}{4}\pi \times (0{\cdot}050)^2\,\text{m}^2$, $p = 70 \times 10^3\,\text{N/m}^2$, $w = 9{\cdot}81 \times 10^3\,\text{N/m}^3$ and $h = 1{\cdot}8\,\text{m}$

$$Q = 0{\cdot}6 \times \tfrac{1}{4}\pi \times 0{\cdot}25 \times 10^{-2} \sqrt{\left[2 \times 9{\cdot}81\left(\frac{70 \times 10^3}{9{\cdot}81 \times 10^3} + 1{\cdot}8\right)\right]}$$

$$= 0{\cdot}0156\,\text{m}^3/\text{s}$$

7.6 Re-entrant orifice

A pipe enters a vessel horizontally, forming a re-entrant sharp-edged orifice of 40 mm diam. If the head above the orifice is 1·5 m and the coefficient of velocity is 0·95, find (a) the coefficient of contraction and the discharge when the orifice is running free, and (b) the discharge when the orifice is running full, assuming that a *vena contracta* is formed in the pipe, the coefficient of contraction being the same as in (a), and allowing for the energy loss between the *vena contracta* and the outlet.

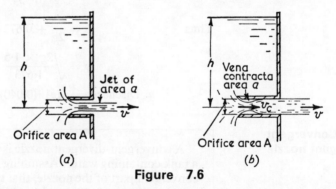

Figure 7.6

Solution. The two cases are shown in Fig. 7.6(a). By Newton's 2nd Law,

Rate of change of momentum of jet = Pressure force on orifice

If Q = volume passing per second

$$\frac{wQv}{g} = wh \times A$$

or, since $Q = av$,

$$\frac{av^2}{g} = hA$$

But, from Example 7.1,

$$v = C_v\sqrt{(2gh)}$$

$$\frac{a}{g} C_v{}^2 2gh = hA$$

$$\text{Coefficient of contraction } C_c = \frac{a}{A} = \frac{gh}{2C_v{}^2 gh} = \frac{1}{2C_v{}^2}$$

$$\text{Putting } C_v = 0{\cdot}95, \quad C_c = \frac{1}{2 \times 0{\cdot}95^2} = \mathbf{0{\cdot}554}$$

$$\text{Discharge } Q = av = 0{\cdot}544 A C_v \sqrt{(2gh)}$$
$$= 0{\cdot}554 \times \tfrac{1}{4}\pi(0{\cdot}040)^2 \times 0{\cdot}95\sqrt{(2g \times 1{\cdot}5)}$$
$$= \mathbf{0{\cdot}0036\,m^3/s}$$

(b) Applying Bernoulli's equation to the free surface and outlet,

$$h = \frac{v^2}{2g} + \text{loss of energy}$$

It will be shown in Example 10.1 that the loss of energy occurring as the stream enlarges from the *vena contracta* to the full pipe area is $(v_c - v)^2/2g$ where v_c = velocity at the *vena contracta*.

For continuity of flow, $a \times v_c = A \times v$

$$\text{Loss of energy} = \frac{v^2}{2g}\left(\frac{A}{a} - 1\right)^2$$

From (a) above $\qquad \dfrac{a}{A} = 0{\cdot}554$

$$\text{Loss of energy} = \frac{v^2}{2g}\left(\frac{1}{0{\cdot}554} - 1\right)^2 = 0{\cdot}673\frac{v^2}{2g}$$

Thus $\qquad h = \dfrac{v^2}{2g} + 0{\cdot}673\dfrac{v^2}{2g} = 1{\cdot}673\dfrac{v^2}{2g}$

$$v = \sqrt{\frac{2g \times 1{\cdot}5}{1{\cdot}673}} = 4{\cdot}19\,m/s$$

$$\text{Discharge} = Av\tfrac{1}{4}\pi(0{\cdot}040)^2 \times 4{\cdot}19 = \mathbf{0{\cdot}0053\,m^3/s}$$

7.7 Convergent-divergent nozzle

A convergent-divergent nozzle is fitted into the vertical side of a tank containing water. Assuming that there are no losses in the convergent part of the nozzle, that the losses in the divergent part are equivalent to $0{\cdot}18$ times the velocity head at exit, and that the minimum absolute pressure at the throat is $2{\cdot}5$ m of water for a barometric pressure of $10{\cdot}4$ m of water, determine the throat and exit diameters of the nozzle to discharge $0{\cdot}0045$ m³/s for a head of $1{\cdot}5$ m above the centreline of the nozzle.

As the head in the tank is allowed to fall, show that the pressure at the throat is a linear function of the head above the nozzle.

Solution. In Fig. 7.7, apply Bernoulli's equation to the free surface and the throat. If p_0 = atmospheric pressure, p_1 = pressure in throat

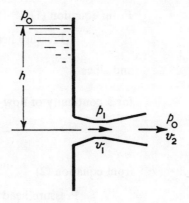

Figure 7.7

and v_1 = velocity in the throat,

$$\frac{p_0}{w} + h = \frac{p_1}{w} + \frac{v_1{}^2}{2g} \qquad . \qquad . \qquad . \qquad . \qquad (1)$$

Putting $p_0/w = 10\cdot4$m and $p_1/w = 2\cdot5$m of water and $h = 1\cdot5$m

$$\frac{v_1{}^2}{2g} = 10\cdot4 - 2\cdot5 + 1\cdot5 = 9\cdot4\text{m of water}$$

$$v_1 = 13\cdot6\text{m/s}$$

Since the discharge is $0\cdot0045\text{m}^3/\text{s}$, if d_1 = throat diameter

$$\tfrac{1}{4}\pi d_1{}^2 v_1 = 0\cdot0045$$

$$d_1{}^2 = \frac{4 \times 0\cdot0045}{\pi \times 13\cdot6}$$

Throat diameter = $d_1 = $ **20·5mm**

Applying Bernoulli's equation to the free surface and exit allowing for the loss in divergence,

$$\frac{p_0}{w} + h = \frac{p_0}{w} + \frac{v_2{}^2}{2g} + 0\cdot18\frac{v_2{}^2}{2g}$$

$$v_2{}^2 = \frac{2gh}{1\cdot18} \qquad . \qquad . \qquad . \qquad . \qquad . \qquad (2)$$

$$v_2 = \sqrt{\frac{2g \times 1\cdot5}{1\cdot18}} = 5\text{m/s}$$

Since discharge is $0\cdot0045\text{m}^3/\text{s}$ if d_2 = exit diameter

$$\tfrac{1}{4}\pi d_2{}^2 v_2 = 0\cdot0045$$

$$d_2{}^2 = \frac{4 \times 0\cdot0045}{\pi \times 5} = 0\cdot00115$$

Throat diameter = $d_2 = 0\cdot034$m = **34mm**

From equation (1)

$$\text{Pressure head at throat} = \frac{p_1}{w} = h + \frac{p_0}{w} - \frac{v_1{}^2}{2g}$$

and since

$$\tfrac{1}{4}\pi d_1{}^2 v_1 = \tfrac{1}{4}\pi d_2{}^2 v_2$$

for a continuity of flow

$$v_1{}^2 = \left(\frac{d_2}{d_1}\right)^4 v_2{}^2$$

$$= \left(\frac{d_2}{d_1}\right)^4 \frac{2gh}{1\cdot 18}$$

from equation (2)

$$\text{Pressure head at throat} = \frac{p_0}{w} + h - \left(\frac{d_2}{d_1}\right)^4 \frac{h}{1\cdot 18}$$

$$= \frac{p_0}{w} + \left\{1 - \left(\frac{d_2}{d_1}\right)^4\right\} \frac{h}{1\cdot 18}$$

which is a linear function of h.

7.8 Pipe-orifice meter

An orifice meter consists of a 100 mm diam orifice in a 250 mm diam pipe, and has a coefficient of discharge of $0\cdot 65$. The pipe conveys oil of sp. gr. $0\cdot 9$ and the pressure difference between the two sides of the orifice plate is measured by a mercury manometer, the leads to the gauge being filled with oil. If the difference in mercury levels in the gauge is 760 mm, calculate the rate of flow of oil in the pipeline.

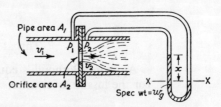

Figure 7.8

Solution. Let v_1, Fig. 7.8, be the velocity and p_1 the pressure immediately upstream of the orifice, and v_2 and p_2 are the corresponding values in the orifice. Then, ignoring losses, by Bernoulli's theorem

$$\frac{v_1{}^2}{2g} + \frac{p_1}{w} = \frac{v_2{}^2}{2g} + \frac{p_2}{w}$$

$$\frac{v_2{}^2 - v_1{}^2}{2g} = \frac{p_1 - p_2}{w}$$

For continuity of flow $A_1 v_1 = A_2 v_2$ or $v_2 = (A_1/A_2)v_1$

$$\frac{v_1{}^2}{2g}\left\{\left(\frac{A_1}{A_2}\right)^2 - 1\right\} = \frac{p_1 - p_2}{w}$$

$$v_1 = \frac{A_2}{\sqrt{(A_1{}^2 - A_2{}^2)}}\sqrt{2g\left(\frac{p_1 - p_2}{w}\right)}$$

Actual discharge = coefficient of discharge × theoretical discharge

$$Q = C_d A_1 v_1 = C_d \frac{A_1 A_2}{\sqrt{(A_1{}^2 - A_2{}^2)}}\sqrt{2g\left(\frac{p_1 - p_2}{w}\right)}$$

or
$$Q = \frac{C_d A_1}{\sqrt{(m^2 - 1)}}\sqrt{2g\left(\frac{p_1 - p_2}{w}\right)}$$

where
$$m = \frac{A_1}{A_2}$$

Considering the U-tube gauge, since pressures are equal at level XX

$$p_1 + wx = p_2 + w_g x$$

$$\frac{p_1 - p_2}{w} = x\left(\frac{w_g}{w} - 1\right)$$

Putting $x = 760\,\text{mm} = 0\cdot76\,\text{m}$ and

$$\frac{w_g}{w} = \frac{13\cdot6}{0\cdot9} = 15\cdot1$$

$$\frac{p_1 - p_2}{w} = 0\cdot76 \times 14\cdot1 = 10\cdot72\,\text{m of oil}$$

$$C_d = 0\cdot65$$

$$A_1 = \tfrac{1}{4}\pi(0\cdot25)^2 = 0\cdot0497\,\text{m}^2$$

$$m = \frac{A_1}{A_2} = \frac{\tfrac{1}{4}\pi \times (0\cdot25)^2}{\tfrac{1}{4}\pi \times (0\cdot10)^2} = 6\cdot25$$

$$m^2 = 39\cdot1$$

$$Q = \frac{0\cdot65 \times 0\cdot0497}{6\cdot17}\sqrt{(2g \times 10\cdot72)}$$

$$= 0\cdot00524 \times 14\cdot5 = \mathbf{0\cdot0762\,m^3/s}$$

7.9 Large orifice

A reservoir discharges through a sluice gate $0\cdot9$ m wide by $1\cdot2$ m deep. The top of the opening is $0\cdot6$ m below the water level in the reservoir and the downstream water level is below the bottom of the opening. Calculate (a) the theoretical discharge through the opening, and (b) the percentage error if the opening is treated as a small orifice.

Solution. The gate forms a large orifice and the velocity of flow will be greater at the bottom than at the top of the opening.

(*a*) Consider a horizontal strip across the opening of thickness δh at a depth h below the free surface (Fig. 7.9):

$$\text{Area of strip} = B\delta h$$
$$\text{Velocity of flow through strip} = \sqrt{(2gh)}$$
$$\text{Discharge through strip} = B\sqrt{(2g)}h^{1/2}\delta h$$

For the whole opening

$$\text{Discharge } Q = B\sqrt{(2g)}\int_{H_1}^{H_2} h^{1/2}dh$$

$$= \tfrac{2}{3}B\sqrt{(2g)}(H_2{}^{3/2} - H_1{}^{3/2})$$

Putting $B = 0.9\,\text{m}$, $H_1 = 0.6\,\text{m}$, $H_2 = 1.8\,\text{m}$

$$Q = \tfrac{2}{3} \times 0.9 \times \sqrt{(2g)}(1.8^{3/2} - 0.6^{3/2})$$

$$= 2.68(2.415 - 0.465) = \mathbf{5.18\,m^3/s}$$

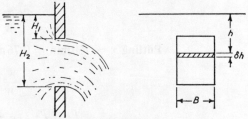

Figure   7.9

(*b*) For a small orifice $Q = A\sqrt{(2gh)}$

where A = area of opening, h = head above centreline. Putting $A = 0.9 \times 1.2 = 1.08\,\text{m}^2$ and $h = 1.2\,\text{m}$

$$Q = 1.08\sqrt{(2 \times 9.81 \times 1.2)} = 5.25\,\text{m}^3/\text{s}$$

$$\text{Error} = \frac{5.24 - 5.18}{5.18} \times 100 = \mathbf{1.16\ per\ cent}$$

Problems

1   A sharp-edged orifice, of 50 mm diam, in the vertical side of a large tank, discharges under a head of 4.8 m. If $C_c = 0.62$ and $C_v = 0.98$, determine (*a*) the diameter of the jet, (*b*) the velocity of the jet at the *vena contracta*, (*c*) the discharge in dm³/s.
Answer 40.3 mm, 9.5 m/s, 12.15 dm³/s

2   Compensation water is to be discharged by two circular orifices under a constant head 0.75 m measured to the centre of the orifice. What diameter will be required to give 13 620 m³ per day? $C_c = 0.62$ and $C_v = 0.97$.
Answer 0.2085 m

3 Define the coefficients for flow from orifices.

A tank $1 \cdot 2$ m square and $3 \cdot 6$ m high receives water at a rate of 20 dm³/s. The water flows from the tank through a 65 mm diam orifice having $C_d = 0 \cdot 79$. Find the height of water above the orifice to maintain flow.

Answer $2 \cdot 97$ m

4 A sharp-edged orifice of $48 \cdot 2$ mm diam is employed to measure the supply of air to an oil engine. Prove that the volume in m³/min passing through the orifice is $0 \cdot 92\sqrt{(h/\rho)}$ where h is the drop of pressure between the two sides of the orifice in cm of water and ρ is the density of the air, assumed uniform, in kg/m³, and the coefficient of discharge for the orifice is $0 \cdot 602$.

Calculate the volume of air in m³/min if $h = 2 \cdot 16$ cm of water, the pressure of the atmosphere 775 mm of mercury and the temperature $15 \cdot 8°C$. For air $pV = 287T$ in SI units.

Answer $1 \cdot 2$ m³/min

5 A closed tank partially filled with water discharges through an orifice of $12 \cdot 5$ mm diam and has a coefficient of discharge of $0 \cdot 7$. If air is pumped into the upper part of the tank, determine the pressure required to produce a discharge of 36 dm³/min when the water surface is $0 \cdot 9$ m above the outlet.

Answer $15 \cdot 7$ kN5m²

6 A reservoir discharges through a 50 mm diam opening into a conical pipe expanding to 100 mm diam. What is the discharge in dm³/s when the reservoir level is 6 m above the opening?

Answer 85 dm³/s

7 A sharp-edged orifice of $12 \cdot 5$ mm diam is situated in the base of an otherwise closed tank. At a given instant the head of water above the orifice is $1 \cdot 8$ m. If the discharge of water is to be 90 kg/min at this instant, find the pressure of air which must be pumped in above the water. $C_d = 0 \cdot 6$.

Answer 190 kN/m²

8 A circular sharp-edged orifice of 25 mm diam is situated in a vertical side of a large oil tank and discharges to a region where the pressure is 83 kN/m² absolute. The oil in the tank is maintained at a constant level, and is subject to an external pressure of 48 kN/m² gauge. The coefficients of velocity and contraction for the orifice are $0 \cdot 98$ and $0 \cdot 62$ respectively, the specific gravity of the oil is $0 \cdot 85$ and the barometric pressure head is 760 mm of mercury abs.

If the jet droops 5 cm in a horizontal distance of 138 cm, calculate (*a*) the height of the oil surface in the tank above the axis of the orifice, (*b*) the volume rate of flow through the orifice.

Answer $1 \cdot 96$ m, $4 \cdot 16$ dm³/s

9 Find the diameter of a circular orifice to discharge 15 dm³/s under a head of $1 \cdot 5$ m using a coefficient of discharge of $0 \cdot 6$. If the orifice is in a vertical plane and the jet falls 250 mm in a horizontal distance of $1 \cdot 2$ m from the *vena contracta*, find the value of the coefficient of contraction.

Answer $76 \cdot 6$ mm, $0 \cdot 612$

10 A large tank has a circular sharp-edged orifice 930 mm² in area at a depth of 2·7 m below constant water level. The jet issues horizontally and in a horizontal distance of 2·34 m it falls 0·54 m. The measured discharge is 4·2 dm³/s. Calculate the coefficients of velocity, contraction and discharge.
Answer 0·97, 0·624, 0·643

11 A tank of square cross-section, each side measuring 0·3 m, is open at the top and is fixed in an upright position. A 6 mm diam circular orifice is situated in one of the vertical sides near the bottom. Water flows into the tank at a constant rate of 280 dm³/h. At a particular instant, the jet strikes the floor at a point 0·63 m from the *vena contracta*, measured horizontally, and 0·53 m below the centreline of the orifice measured vertically. Determine whether the water surface in the tank is rising or falling at the instant under consideration. Also find the height of the surface above the centreline of the orifice and the rate of change of height. (Take C_v as 0·97 and C_d as 0·64.)
Answer 0·197 m, 0·46 mm/s

12 The velocity of the jet from an orifice in the vertical side of a tank may be determined from the position of three points p_0, p_1 and p_2 in the path of the jet. If p_1 and p_2 are distances x_1 and x_2 to the right and y_1 and y_2 below p_0 respectively, obtain an expression for the velocity of discharge.

Answer $$\left(v = \sqrt{\frac{gx_1x_2(x_1 - x_2)}{2(x_2y_1 - x_1y_2)}} \right)$$

13 A sharp-edged orifice 50 mm in diam in the vertical side of a tank discharges under a head of 3 m. If $C_c = 0·62$, and $C_v = 0·98$, how far from the vertical plane containing the orifice will the jet strike a horizontal plane 1·8 m below the plane of the centre of the orifice?
Answer 4·56 m

14 A circular orifice of area 6·5 cm² is made in the vertical side of a large tank. The tank is suspended from knife edges 1·5 m above the level of the orifice. When the head of water is 1·2 m the discharge is 118·5 kg/min and a turning moment of 14·4 N-m has to be applied to the knife edges to keep the tank vertical. Determine (*a*) the coefficient of velocity, (*b*) the coefficient of contraction, (*c*) the coefficient of discharge of the orifice.
Answer 0·98, 0·627, 0·614

15 A vertical triangular orifice in the wall of a reservoir has a base 0·9 m long, 0·6 m below its vertex and 1·2 m below the water surface. Determine the theoretical discharge.
Answer 1·19 m³/s

16 Calculate the theoretical discharge through a vertical orifice 0·3 m square in the side of a tank if the free surface of the liquid in the tank is 150 mm above the top edge of the orifice. Compare this value with that obtained by multiplying the orifice area by the theoretical velocity at its centre.
Answer 0·216 m³/s, 0·218 m³/s

17 A Borda mouthpiece of 150 mm diam discharges water under a head of 3 m. Calculate (*a*) the discharge and (*b*) the diameter of the jet at the *vena contracta* when running free.
Answer   67·8 dm³/s, 106 mm

18 A convergent-divergent nozzle is fitted into the side of a tank containing water and, under a constant head, *H*m above the centreline of the nozzle, discharges into the atmosphere. Obtain an expression for the ratio of the exit area of the nozzle to the area of the throat for maximum discharge, making the following assumptions: (i) the height of water in the barometer is 10·2 m, (ii) separation will occur at an absolute pressure head of 1·8 m of water, (iii) the only hydraulic losses occur in the divergent portion of the nozzle and amount to 25 per cent of the head lost at a sudden enlargement for the same change of area.

If the supply head *H* is 2·7 m and the throat diam is 50 mm, calculate the maximum discharge.

$$\textit{Answer} \quad \frac{4\sqrt{(H^2 + 6\cdot3H - 17\cdot6)} - (H + 8\cdot4)}{(3H - 8\cdot4)}, \text{ 29 dm}^3/\text{s}$$

19 A 75 mm diam pipeline discharges water through a 50 mm diam orifice. If the pressure in the pipe is 690 kN/m² when the discharge is against atmospheric pressure, what is the theoretical discharge, (*a*) neglecting the velocity in the pipe, (*b*) considering all velocities?
Answer   0·0729 m³/s, 0·0769 m³/s

20 An orifice meter is installed in a vertical pipe to measure the flow of oil, of relative density 0·9, upwards through the pipe. The diameter of the pipe D is 100 mm and that of the orifice is 40 mm. The pressure tappings are at distances *D* upstream and *D*/2 downstream and the pressure difference between them is 8 820 N/m². The coefficient of discharge is 0·60. Starting from Bernoulli's equation find the volume rate of flow along the pipe.
Answer   3·125 dm³/s

8

Flow measurement – notches and weirs

A notch is an opening in the side of a measuring tank or reservoir extending above the free surface and of any suitable geometrical form. It is in effect a large orifice which has no upper edge so that it has a variable area depending on the level of the free surface. A weir is a notch on a large scale, used for example to measure the flow of a river, and may be sharp-edged or have a substantial breadth in the direction of flow.

Rectangular notch

8.1 Rectangular notch theory

Show from first principles that the theoretical rate of flow through a rectangular notch is given by $Q = \frac{2}{3}B\sqrt{(2g)}H^{3/2}$ where B = width of notch and H = height of the water level above the bottom of the notch. Explain why this expression requires modification in practice.

The discharge over a rectangular notch is to be $0 \cdot 14$ m³/s when the water level is 23 cm above the sill. If the coefficient of discharge is $0 \cdot 6$ calculate the width of notch required.

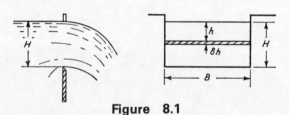

Figure 8.1

Solution. By Torricelli's theorem (Example 7.1) the velocity of a particle discharged at any level h is $\sqrt{(2gh)}$ and will therefore vary from top to bottom of the notch. Considering a horizontal strip (Fig. 8.1) at depth h and of thickness δh

$$\text{Velocity through strip} = \sqrt{(2gh)}$$
$$\text{Area of strip} = B\delta h$$
$$\text{Discharge through strip} = \text{velocity} \times \text{area} = \sqrt{(2gh)} \times B\delta h$$

Integrating from top to bottom of the notch:

$$\text{Total discharge } Q = B\sqrt{(2g)}\int_0^H h^{1/2}dh$$

$$= \tfrac{2}{3}B\sqrt{(2g)}H^{3/2} \qquad . \qquad . \qquad . \quad (1)$$

This is the theoretical discharge through a rectangular notch.

The value of Q given by equation (1) is too high because no account has been taken of energy lost and also because, as shown in Fig. 8.2,

Figure 8.2

there will be a substantial reduction in the width and depth of the notch cross-section because of the curved path lines of the liquid. The actual discharge will be the theoretical discharge multiplied by a coefficient of discharge C_d, so that

$$Q = \tfrac{2}{3}C_d B\sqrt{(2g)}H^{3/2}$$

Putting $Q = 0.14\,\text{m}^3/\text{s}$, $C_d = 0.6$, $H = 23\,\text{cm} = 0.23\,\text{m}$

$$B = \frac{Q}{\tfrac{2}{3}C_d\sqrt{(2g)}H^{3/2}} = \frac{0.14}{\tfrac{2}{3} \times 0.6\sqrt{(2g)}(0.23)^{3/2}}$$

$$= \mathbf{0.72\,m}$$

8.2 Francis formula

State the Francis formula for a rectangular notch or weir, and explain why it is preferable to the simple theoretical formula.

A weir is 6 m long and is situated centrally in a channel 9 m wide. If the head above the sill is 25 cm calculate the discharge.

Solution. If L is the length of the weir and H the head over the sill, the Francis formula for the discharge is

$$Q = 0.623(L - 0.1nH) \times \tfrac{2}{3}\sqrt{(2g)}H^{3/2}$$

or in SI units

$$Q = 1.84(L - 0.1nH)H^{3/2}$$

where n = number of end contractions.

This formula allows for the contraction in width of the weir, due to the curved path of the liquid at the sides, which is found experimentally to be equivalent to $0.1H$ for each side. Fig. 8.3 shows the value of n for typical cases. If the simple theory were used the coefficient of discharge would vary as the value of H varied. The Francis formula is therefore preferable since the correction to the effective width results in a constant coefficient of discharge.

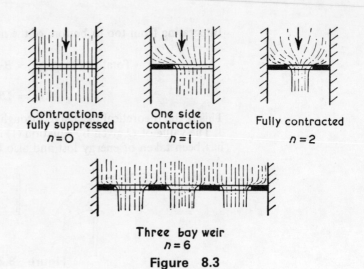

Contractions
fully suppressed
$n = 0$

One side
contraction
$n = 1$

Fully contracted
$n = 2$

Three bay weir
$n = 6$

Figure 8.3

Since $\qquad Q = 1 \cdot 84(L - 0 \cdot 1nH)H^{3/2}$

putting $L = 6\,\text{m}$, $n = 2$ and $H = 25\,\text{cm} = 0 \cdot 25\,\text{m}$

$$Q = 1 \cdot 84(6 - 0 \cdot 2 \times 0 \cdot 25)(0 \cdot 25)^{3/2}$$
$$= 1 \cdot 84(6 - 0 \cdot 05) \times 0 \cdot 125 = \mathbf{1 \cdot 365\,m^3/s}$$

8.3 Bazin and Rehbock formulae

Give an explanation of why the coefficient of discharge of a rectangular notch varies with the head over the sill and state two formulae for the coefficient in terms of the head.

The head over a rectangular weir 12 m wide is 38 cm, and the sill of the weir is $1 \cdot 5$ m above the bed level of the channel. Using Bazin's formula for the coefficient of discharge, calculate the rate of flow.

Solution: If there are full contractions at either side of the notch the effective width of the notch will not remain constant but will vary with the depth. Also as the depth increases the cross-section of the stream will change its shape, as defined by the ratio of width to depth, and the relation between the wetted perimeter of the notch and the head and area of cross-section of stream varies also. The coefficient of discharge will therefore vary with head. The value of the coefficient C_d in the formula

$$Q = \tfrac{2}{3}C_d B\sqrt{(2g)}H^{3/2} \qquad . \qquad . \qquad . \qquad . \qquad (1)$$

can be estimated from

1. *Bazin's formula*:

$$C_d = \left(0 \cdot 607 + \frac{0 \cdot 00451}{H}\right)\left[1 + 0 \cdot 55\left(\frac{H}{P + H}\right)^2\right]$$

where H = head over sill in metres. P = height of sill above floor of channel in metres.

If the correction for P is omitted, equation (1) can be written

$$Q = m\sqrt{(2g)}BH^{3/2}$$

where $m = 0{\cdot}405 + 0{\cdot}003/H$.

2. *Rehbock formula*:

$$C_d = 0{\cdot}605 + \frac{1}{1048H - 3} + \frac{0{\cdot}08H}{P}$$

where H and P are in SI units and the notch extends across the full width of the channel with no end contractions.

Putting $H = 38\text{cm} = 0{\cdot}38\text{m}$ and $P = 1{\cdot}5\text{m}$, Bazin's formula gives

$$C_d = \left(0{\cdot}607 + \frac{0{\cdot}00451}{0{\cdot}38}\right)\left[1 + 0{\cdot}55\left(\frac{0{\cdot}38}{1{\cdot}88}\right)^2\right]$$

$$= 0{\cdot}619(1 + 0{\cdot}025) = 0{\cdot}635$$

$$Q = \tfrac{2}{3}C_d B\sqrt{(2g)}H^{3/2}$$

$$= \tfrac{2}{3} \times 0{\cdot}635 \times 12\sqrt{(2g)}(0{\cdot}38)^{3/2}$$

$$= \mathbf{5{\cdot}26\,m^3/s}$$

8.4 Velocity of approach

Show that a correction factor to allow for velocity of approach to a rectangular notch can be expressed in the form

$$\left[\left(1 + \frac{\alpha h'}{H}\right)^{3/2} - \left(\frac{\alpha h'}{H}\right)^{3/2}\right]$$

where H is the measured head, $h' = v^2/2g$ where v is the velocity of approach, and α is a factor to allow for the velocity not being uniform over the section of the channel.

A rectangular channel $1{\cdot}2$ m wide has at the end a rectangular sharp-edged notch $0{\cdot}9$ m wide, with the sill $0{\cdot}2$ m from the bottom. Assuming $\alpha = 1{\cdot}1$, calculate the flow when the head is 250 mm, using the Francis formula.

Solution. In Example 8.1 it was assumed that the velocity with which the liquid approached the notch was zero, so that the velocity through any strip depended only on its depth below the surface. In fact the stream approaches with a velocity of approach v so that the head producing flow is increased by $\alpha v^2/2g$. Thus if $x = $ head producing flow through a strip at depth h below surface (Fig. 8.1),

$$x = h + \frac{\alpha v^2}{2g} = h + \alpha h'$$

and $\delta h = \delta x$

$$\text{Discharge through strip} = \sqrt{(2gx)} \times B\delta x$$

At the surface $h = 0$ and $x = \alpha h'$, while at the sill $h = H$ and $x = H + \alpha h'$. Integrating between these limits

$$Q = B\sqrt{(2g)}\int_{\alpha h'}^{H+\alpha h'} x^{1/2}dx = \tfrac{2}{3}B\sqrt{(2g)}[(H + \alpha h')^{3/2} - (\alpha h')^{3/2}]$$

$$= \tfrac{2}{3}B\sqrt{(2g)}H^{3/2}\left[\left(1 + \frac{\alpha h'}{H}\right)^{3/2} - \left(\frac{\alpha h'}{H}\right)^{3/2}\right]$$

Comparing this with Example 8.1, equation (1):

$$\text{Correction for velocity of approach} = \left[\left(1 + \frac{\alpha h'}{H}\right)^{3/2} - \left(\frac{\alpha h'}{H}\right)^{3/2}\right]$$

With the correction factor for velocity of approach the Francis formula becomes in SI units

$$Q = 1.84(B - 0.1nH)H^{3/2}\left[\left(1 + \frac{\alpha h'}{H}\right)^{3/2} - \left(\frac{\alpha h'}{H}\right)^{3/2}\right]$$

There are two side contractions giving $n = 2$, $B = 0.9$ m, $H = 0.25$ m and $\alpha = 1.1$.

$$Q = 1.84(0.9 - 0.05)(0.25)^{3/2}\left[\left(1 + \frac{1.1h'}{0.25}\right)^{3/2} - \left(\frac{1.1h'}{0.25}\right)^{3/2}\right]$$

$$= 0.196\{(1 + 4.4h')^{3/2} - (4.4h')^{3/2}\}$$

Also $\qquad\qquad\qquad h' = \dfrac{v^2}{2g}$ and $v = \dfrac{Q}{A}$

where $A = $ area of channel

$$h' = \frac{Q^2}{2gA^2} = \frac{Q^2}{2g(1.2 \times 0.45)^2} = \frac{Q^2}{5.73}$$

The solution for Q is most readily found by successive approximation. If $h' = 0$, $Q = 0.196$ m³/s.

Taking this value of Q we have

$$h' = \frac{(0.196)^2}{5.73} = 0.0067\,\text{m}$$

Putting $h' = 0.0067$,

$$Q = 0.196\{(1.029)^{3/2} - (0.029)^{3/2}\} = 0.203\,\text{m}^3/\text{s}$$

and for this value of Q we have

$$h' = \frac{(0.203)^2}{5.73} = 0.0072\,\text{m}$$

Putting $h' = 0.0072$,

$$Q = 0.196\{(1.032)^{3/2} - (0.032)^{3/2}\} = \mathbf{0.204\,m^3/s}$$

V-notches

8.5 V-notch theory

Develop an expression for the quantity of liquid flowing over a sharp-edged V-notch of total angle 2θ, in terms of the head H above the bottom of the notch, the angle θ, and the coefficient of discharge C_d, assuming the velocity of approach to be small.

If the rate of flow of water over a V-notch having $\theta = 35°$ is $42 \cdot 5$ dm³/s, find the head in centimetres. Take C_d as $0 \cdot 62$.

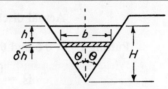

Figure 8.4

Solution. Since the velocity of flow through the notch varies from top to bottom, consider a strip of thickness δh (Fig. 8.4) at a depth h below the surface. If the velocity of approach is small:

$$\text{Head producing flow} = h$$

$$\text{Velocity through strip} = v = \sqrt{(2gh)}$$

$$\text{If width of strip} = b, \quad \text{Area of strip} = b\delta h$$

$$\text{Discharge through strip} = \delta Q = vb\delta h$$

The width b depends on h and is given by

$$b = 2(H - h) \tan \theta$$

Thus
$$\delta Q = \sqrt{(2g)}h^{1/2} \times 2(H - h) \tan \theta \times \delta h$$
$$= 2\sqrt{(2g)} \tan \theta (Hh^{1/2} - h^{3/2})\delta h$$

Integrating between the limits $h = 0$ and $h = H$

$$Q = 2\sqrt{(2g)} \tan \theta \int_0^H (Hh^{1/2} - h^{3/2})dh$$

$$= 2\sqrt{(2g)} \tan \theta \left[\frac{2}{3} Hh^{3/2} - \frac{2}{5} h^{5/2} \right]_0^H$$

$$Q = \frac{8}{15} \sqrt{(2g)} \tan \theta \cdot H^{5/2}$$

This is the theoretical discharge.

$$\text{Actual discharge} = C_d \times Q = C_d \frac{8}{15} \sqrt{(2g)} \tan \theta \cdot H^{5/2}$$

Putting $C_d = 0 \cdot 62$, $\theta = 35°$, $\tan \theta = 0 \cdot 70021$

$$\text{Actual discharge} = 0 \cdot 0425 \text{m}^3/\text{s}$$

$$0 \cdot 0425 = 0 \cdot 62 \times \frac{8}{15} \sqrt{(2g)} \times 0 \cdot 70021 \times H^{5/2}$$

$$H^{5/2} = 0 \cdot 0414$$

$$H = 0 \cdot 28 \text{m} = \mathbf{28\,cm}$$

8.6 Errors

Develop, from first principles, a discharge formula of the form $Q = Kh^n$, giving the numerical values of K and n, for a V-notch having a total angle (2θ) to 60 deg and $C_d = 0.6$, if Q is in m³/s and h is in metres.

If an error of 1·8 mm is made in reading a head of 180 mm, what percentage error will result in the value of Q?

Solution. From Example 8.5,

$$\text{Actual discharge } Q = C_d \frac{8}{15} \sqrt{(2g)} \tan \theta \,.\, H^{5/2}$$

Putting $C_d = 0.6$ and $\theta = \frac{1}{2} \times 60 = 30°$

$$Q = \frac{0.6 \times 8\sqrt{(2g)}}{15} \tan 30° H^{5/2}$$

$$= 1.42 \times 0.5774 H^{5/2} = 0.82 H^{5/2}$$

This equation gives Q in m³/s when H is in metres.

Thus $\qquad\qquad K = \textbf{0·82} \quad$ and $\quad n = \textbf{2·5}$

$$Q = 0.82h^{2.5} \quad \text{and} \quad \delta Q = 2.5 \times 0.82h^{1.5}\delta h$$

$$\delta Q / Q = 2.5\delta h / h$$

Putting $\delta h = 1.8$ mm when $h = 180$ mm

$$\frac{\delta Q}{Q} = 2.5 \times \frac{1.8}{180} = \frac{2.5}{100}$$

Percentage error in $Q = \textbf{2·5 per cent}$

8.7 Trapezoidal notch

A trapezoidal notch has a base of width L and each side makes an angle of θ to the vertical. Deduce an expression for the discharge through the notch.

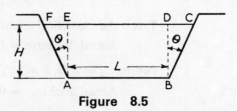

Figure 8.5

Solution. Fig. 8.5 shows the notch which can be considered as consisting of a rectangular notch ABDE of width L and two triangular areas AEF and BDC which together form a V-notch of angle 2θ.

From Example 8.1,

Theoretical discharge through rectangular portion

$$= \tfrac{2}{3}L\sqrt{(2g)}H^{3/2}$$

From Example 8.5,

Theoretical discharge through triangular portion

$$= \frac{8}{15}\sqrt{(2g)}\tan\theta \, . \, H^{5/2}$$

$$\text{Total theoretical discharge} = \frac{2}{3}\sqrt{(2g)}LH^{3/2} + \frac{8}{15}\sqrt{(2g)}\tan\theta \, . \, H^{5/2}$$

$$= \frac{2}{3}\sqrt{(2g)}H^{3/2}\left(L + \frac{4}{5}\tan\theta \, . \, H\right)$$

$$\text{Actual discharge} = C_d\frac{2}{3}\sqrt{(2g)}H^{3/2}\left(L + \frac{4}{5}\tan\theta \, . \, H\right)$$

8.8 Cipolletti weir

> What is a Cipolletti weir and what are its advantages?
> A fully-contracted trapezoidal notch has a coefficient of discharge of $0 \cdot 623$ and a base width L. Calculate the angle of inclination θ of the sides of the notch if it is to have the same discharge under any head H as that given by the Francis formula for a rectangular notch of width L with no side contractions.

Solution. The Cipolletti weir is a fully-contracted trapezoidal notch so designed that the flow through the triangular portion is equal to the loss of flow due to side contractions on a rectangular notch of the same base width. The Cipolletti weir thus functions as if it were a rectangular notch without end contractions.

The Francis formula for a fully-contracted notch is

$$Q = 1 \cdot 84(L - 0 \cdot 2H)H^{3/2}$$

If there are no end contractions

$$Q = 1 \cdot 84LH^{3/2}$$

Thus, loss of discharge due to end contractions $= 1 \cdot 84 \times 0 \cdot 2H^{5/2}$.

If the Cipolletti weir is to function as if it were a rectangular notch without end contractions, referring to Fig. 8.5,

Additional flow through triangular portions AFE and BDC

$$= \text{loss of discharge due to end contractions}$$

From Example 8.5,

Flow through triangular portion $= C_d \times \frac{8}{15}\sqrt{(2g)}\tan\theta \, . \, H^{5/2}$

so that $\quad 0 \cdot 623 \times \dfrac{8}{15}\sqrt{(2g)}\tan\theta \, . \, H^{5/2} = 1 \cdot 84 \times 0 \cdot 2H^{5/2}$

$$\tan\theta = \frac{1 \cdot 84 \times 0 \cdot 2 \times 15}{0 \cdot 623 \times 8 \times \sqrt{(2g)}} = \frac{1}{4}$$

$$\boldsymbol{\theta = 14° \, 2'}$$

8.9 Effect of nappe conditions

Describe, with sketches, what is meant by the terms "depressed nappe," "drowned nappe" and "clinging nappe" as applied to the discharge side of a weir. How do these conditions affect the discharge over a weir?

Solution. For accurate measurement of flow the stream passing over the weir should spring clear as in Fig. 8.6(*a*) and the space under the nappe should be at atmospheric pressure and fully ventilated.

Depressed nappe is shown in Fig. 8.6(*b*). The space between the weir and the nappe is only partially ventilated so that the nappe does not spring clear. Discharge is about 8 to 10 per cent greater than with a free nappe.

Drowned nappe. The downstream water level is above the crest of the weir as shown in Fig. 8.6(*c*). The discharge depends on the ratio of H_1 to H_2 and can be calculated from the formula $Q = 1 \cdot 84 K L H_1^{3/2}$ in SI

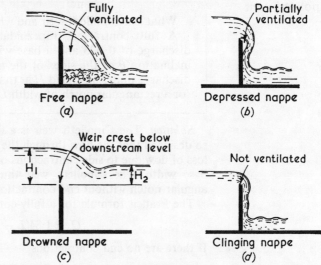

Figure 8.6

units where K is a constant and L the breadth of the weir provided the ratio H_2/H_1 does not exceed 0·22. Difficulties arise from the formation of waves downstream.

Clinging nappe is shown in Fig. 8.6(*d*). There is no ventilation of the nappe which adheres to the face of the weir. Discharge is increased by 20 to 30 per cent above that with a free nappe.

Problems

1 Using the Francis formula, determine the length of the sill of the waste weir of a storage reservoir if the maximum daily flow over the weir is 125 000 m³ and the level of the water must not rise more than 0·6 m above the sill. Assume that the weir is fully contracted.

Answer 1·81 m

2 Using Francis's formula, determine the discharge over a rectangular notch 0·9 m long with heads of 75 mm and 150 mm (*a*) with no side contractions, (*b*) with one side contraction, (*c*) with two side contractions.

Answer 34, 96, 33·7, 94·4, 33·4, 92·6 dm³/s

3 A weir has a length of 9 m and Bazin's coefficient is 0·415. It is required to discharge 5·72 m³/s. Calculate the head above the sill.

Answer 0·492 m

4 What is the discharge over a weir 275 m long if Bazin's coefficient $m = 0·405 + (0·003/h)$ and the head is 1·2 m?

Answer 652·6 m³/s

5 A rectangular notch has a width of 0·9 m, the head over the notch is 0·2 m and the height of the sill above the bed level is 0·15 m. If the breadth of the channel is 1·2 m, calculate the discharge in dm³/s taking into account the velocity of approach. Use the Francis formula suitably modified.

Answer 148 dm³/s

6 Working from first principles, and taking $C_d = 0·623$, develop the Francis formula for flow of water over a sharp-edged suppressed rectangular weir with a velocity of approach in the form

$$Q = 1·84B\,[(H + h)^{3/2} - h^{3/2}]$$

where B = breadth of weir, H = measured head and h = kinetic head of approach, all in metric units.

Such a weir is 0·75 m wide with a head of 0·38 m and the crest is 0·6 m above the bed. Making reasonable approximations estimate the discharge of water.

Answer 0·334 m³/s

7 Experiments on the flow over a rectangular notch 0·9 m wide placed in a channel 1·2 m wide with the sill 0·6 m above the bed of the channel give the following results –

| Q, m³/s | 0·273 | 0·510 | 0·794 |
| h, m | 0·297 | 0·448 | 0·597 |

Show that these observations are consistent with the formula $Q = KbH^n$ if $H = h + (v^2/2g)$ where v is the velocity of flow in the channel. Determine the values of K and n.

Answer 1·84, 1·5

8 A sharp-crested weir 0·9 m high extends fully across a rectangular channel 6 m wide in which there is 2·83 m³/s flowing. Determine the depth of water upstream from the weir by the Francis formula.

Answer 1·30 m

9 Establish the formula for the flow of water over a V-notch in terms of the head. Use this formula to calculate the quantity of water flowing over a V-notch having an angle of 90 deg when the head is $0 \cdot 2$ m. Assume a coefficient of discharge of $0 \cdot 6$.
Answer $25 \cdot 3$ dm³/s

10 Find the depth and top width of a triangular notch capable of discharging maximum quantity of $0 \cdot 7$ m³/s and such that the head shall be 75 mm when the discharge is $5 \cdot 6$ dm³/s.
Answer $0 \cdot 517$ m, $1 \cdot 591$ m

11 Working from first principles, obtain an expression for the rate of flow over a V-notch in terms of the angle θ between the sides and the head H above the bottom of the notch.

Calculate the value of the angle θ for a V-notch which is to discharge $0 \cdot 41$ m³/s under a head of $0 \cdot 6$ m, assuming a coefficient of discharge of $0 \cdot 6$.
Answer $91°24'$

12 Develop a formula for the theoretical discharge over a V-notch.

Explain the way in which this may be modified to give the actual discharge, giving reasons why the theoretical discharge does not agree with the actual discharge.

A 90-deg V-notch has a coefficient of discharge of $0 \cdot 6$. Calculate the discharge when the observed head is $0 \cdot 65$ m above the bottom of the V.
Answer $0 \cdot 472$ m²

13 Develop the formula $C_d \times \frac{8}{15} \tan \theta \sqrt{(2g)} H^{5/2}$ for the flow in m³/s over a V-notch having an included angle of 2θ.

Find the flow if the measured head is $0 \cdot 375$ m, θ being 45 deg and C_d being $0 \cdot 6$. If this flow is wanted within an accuracy of 2 per cent up or down, what are the limiting values of the head?
Answer $0 \cdot 122$ m³/s; $0 \cdot 372$ m, $0 \cdot 378$ m

14 When water flows through a right-angled V-notch, show that the discharge is given by $Q = KH^{5/2}$ in which K is a constant and H is the height of the surface of the water above the bottom of the notch. If H is measured in centimetres and the coefficient of contraction is $0 \cdot 6$, what is the value of K?

A V-notch is fitted to the end of a large tank and is used for measuring the discharge of water from the condenser of a small steam engine. Describe, with the help of a sketch, a gauge for measuring H, and explain how the zero of such a gauge would be obtained.
Answer $1 \cdot 418 \times 10^{-5}$

15 Derive an expression for the quantity of flow over a V-notch of angle 2θ when the head over the apex of the notch is h.

In an experiment on a 90° V-notch the flow is collected in a $0 \cdot 9$ m diam vertical cylindrical tank. It is found that the depth of water in the tank increases by $0 \cdot 685$ m in $16 \cdot 8$ s when the head over the notch is $0 \cdot 2$ m. Determine the coefficient of discharge of the notch.
Answer $0 \cdot 615$

16 A 90° V-notch and a rectangular weir are placed in series. The length of the rectangular weir is 0·6 m and its coefficient is 1·81. If the coefficient of contraction of the V-notch is 0·61, what will be its working head when the head on the weir is 0·15 m?
Answer   0·289 m

17 A channel is conveying 0·28 m³/s of water. Assuming that an error of 1·5 mm may be made in measuring the head, determine the percentage error resulting (*a*) from the use of a right-angled triangular weir with $C_d = 0·6$, (*b*) from the use of a rectangular weir 3 m long with fully suppressed end contractions (Francis formula)
Answer   (*a*) 0·72 per cent; (*b*) 1·64 per cent

18 What is the theoretical rate of flow in m³/s from the notch shown (Fig. 8.7), if the total head above the sill of the notch is 0·15 m?
Answer   0·0403 m³/s

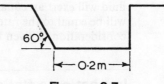

60°

|← 0·2 m →|

Figure 8.7

19 A trapezoidal notch of the form devised by Cipolletti has a base width of 1·8 m and a coefficient of discharge of 0·623. Using the Francis formula for a rectangular notch $Q = 1·84[B - (nH/10)]H^{3/2}$, find the angle of slope of the sides of the notch for a constant coefficient of discharge and hence obtain an equation of flow for the Cipolletti notch of the form $Q = KH^{3/2}$.

If the still water head over the notch is 0·38 m, what is the flow in cubic metres per second?
Answer   0·78 m³/s

20 The breadth of a Cipolletti weir is 0·5 m, the measured head 0·21 m, with a velocity of approach of 1·5 m/s. Calculate the discharge if the weir coefficient is 1·84.
Answer   0·135 m³/s

9

Force exerted by a jet

Newton's laws of motion can be stated as follows:

1. A body will remain in the same condition of rest or of motion with uniform velocity in a straight line until acted upon by an external force.

2. The rate of change of momentum of a body is proportional to the force acting upon it and takes place in the line of action of the force.

3. To every action there is an equal and opposite reaction.

Thus whenever a stream of fluid strikes a surface, or is deflected from its path for example by a curved vane, or has its velocity changed either in magnitude or direction, there must be an external force acting and the fluid will exert an equal and opposite force. The magnitude of the force will be equal to the rate of change of momentum of the stream if proper consideration is given to the correct use of units.

Impact on a flat surface

9.1 Normal impact on stationary plate

Develop an expression for the force exerted by a jet of liquid which strikes a stationary flat plate normally.

A jet of water from a 50 mm diam nozzle impinges normally with a velocity of 6·3 m/s on a stationary flat plate. Calculate the force exerted by the water on the plate.

Solution. In Fig. 9.1 the velocity of the jet $= v$, the cross-sectional area $= a$ and the mass density $= \rho$. The jet strikes the plate and does not rebound but spreads sideways over the surface of the plate. The momentum normal to the plate is destroyed.

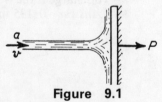

Figure 9.1

Volume of water striking plate per second $= av$

Mass of water striking plate per second $= \rho av$

Initial velocity normal to plate $= v$

Final velocity on impact $= 0$

Change of velocity $= v$

Force exerted on plate = rate of change of momentum of jet

= mass per sec × change of velocity

$$P = \rho av \times v = \rho av^2$$

Putting $\rho = 10^3 \text{kg/m}^3$, $a = \frac{1}{4}\pi(0{\cdot}05)^2\text{m}^2$, $v = 6{\cdot}3\text{m/s}$

$$P = 10^3 \times \tfrac{1}{4}\pi(0{\cdot}05)^2 \times 6{\cdot}3^2 = 78\text{N}$$

9.2 Normal impact on single moving plate

> A jet of water issues from a nozzle with a velocity v and strikes normally a flat plate which is moving with a velocity u in the same direction as the jet. If the cross-sectional area of the jet is a, find the force exerted on the plate by the jet.
>
> A jet of water of 22·5 cm diam impinges normally on a flat plate moving at 0·6 m/s in the same direction as the jet. If the discharge is 0·14 m³/s, find the force and the work done per second on the plate.

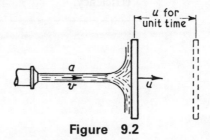

Figure 9.2

Solution. From Fig. 9.2 it can be seen that the length of the jet is continually increasing. Part of the fluid leaving the nozzle is required to extend the length of the jet, thus reducing the amount that strikes the plate.

Mass of fluid leaving nozzle/sec $= \rho av$

In unit time the plate moves a distance u increasing the volume of fluid in the jet by au.

Mass of fluid used to extend jet/sec $= \rho au$

Mass of fluid striking plate/sec $= \rho a(v - u)$

Initial velocity of jet $= v$

Final velocity of jet = velocity of plate $= u$

Change of velocity $= v - u$

Force on plate = rate of change of momentum of jet

= mass/sec × change of velocity

$$P = \rho a(v - u) \times (v - u) = \rho a(v - u)^2$$

Putting $\rho = 1000\,\text{kg/m}^3$,

$$a = \tfrac{1}{4}\pi(0{\cdot}225)^2 = 0{\cdot}0398\,\text{m}^2$$

$$v = \frac{Q}{a} = \frac{0 \cdot 140}{0 \cdot 0398} = 3 \cdot 52 \,\text{m/s}$$

$$u = 0 \cdot 6 \,\text{m/s}$$

Force on plate $= P = 1000 \times 0 \cdot 0398(3 \cdot 52 - 0 \cdot 6)^2 = \textbf{339N}$

Work done on plate/sec $=$ force exerted $\times$ distance moved/sec

$$= P \times u$$

$$= 339 \times 0 \cdot 6 = 204 \,\text{N-m/s or J/s}$$

$$= \textbf{204 W}$$

9.3 Series of flat plates

In an undershot waterwheel the cross-sectional area a of the stream striking the series of radial flat vanes of the wheel is $0 \cdot 1$ m² and the velocity v of the stream is 6 m/s. The velocity u of the vanes is 3 m/s. Calculate the force P exerted on the series of vanes by the stream, the work done per second, and the hydraulic efficiency.

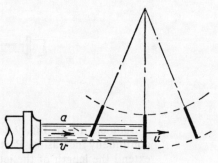

Figure   9.3

Solution. Since there are a series of vanes on the wheel (Fig. 9.3) the average length of the jet from the nozzle to the point of impact remains constant and all the water from the nozzle strikes one or other of the vanes. Assuming that the diameter of wheel is large so that impact is approximately normal,

Mass of water striking vanes/sec $= \rho a v$

Initial velocity of water $= v$

Final velocity $=$ velocity of vane $= u$

Change of velocity on impact $= v - u$

Force of water on vane $=$ rate of change of momentum of jet

$$= \text{mass/sec} \times \text{change of velocity}$$

$$P = \rho a v(v - u)$$

Putting $\rho = 1000 \,\text{kg/m}^3$, $a = 0 \cdot 1 \,\text{m}^2$, $v = 6 \,\text{m/s}$, $u = 3 \,\text{m/s}$

Force on vanes $= P = 1000 \times 0 \cdot 1 \times 6 \times 3 = \textbf{1800N}$

$$\text{Work done on vanes/sec} = \text{force} \times \text{distance moved/sec}$$
$$= P \times u$$
$$= 1800 \times 3$$
$$= \textbf{5400 N-m/s or W}$$

$$\text{Efficiency} = \frac{\text{Work done}}{\text{Energy supplied}}$$

The energy supplied is the kinetic energy of the jet per second.

Kinetic energy of jet per second

$$= \tfrac{1}{2}\rho a v^3$$
$$= 1000 \times 0\cdot1 \times (\tfrac{1}{2} \times 6^3) = \textbf{10800 W}$$

$$\text{Efficiency} = \frac{5400}{10\,800} = 0\cdot5 = \textbf{50 per cent}$$

9.4 Inclined plate

Derive an expression for the force exerted by a jet of area a which strikes a flat plate at an angle θ to the normal to the plate with a velocity v if (a) the plate is stationary, (b) the plate is moving in the direction of the jet with a velocity u.

A jet of water 25 mm in diam, moving with a velocity of 6 m/s, strikes a flat plate at an angle of 30 deg to the normal to the plate. If the plate itself is moving with a velocity of $1\cdot5$ m/s and in the direction of the normal to its surface, calculate the normal force exerted on the plate. Find also the work done per sec and the efficiency.

Solution. When the jet strikes the inclined plate (Fig. 9.4(a)) it does not rebound but flows out over the plate in all directions. If the plate is smooth and frictionless the flow divides so that the momentum parallel to the plate is unaltered. The momentum normal to the plate is destroyed on impact.

Normal force on plate

$\quad$ = rate of change of momentum normal to plate

$\quad$ = mass/sec striking plate × change of velocity normal to plate

(a) $\qquad$ Mass striking stationary plate/sec = $\rho a v$

Initial component velocity of jet normal to plate = $v \cos \theta$

$$\text{Final velocity after impact} = 0$$
$$\text{Change of velocity} = v \cos \theta$$

Force exerted normal to plate

$$= \text{mass/sec} \times \text{change of velocity normal to plate}$$
$$= \rho a v^2 \cos \theta$$

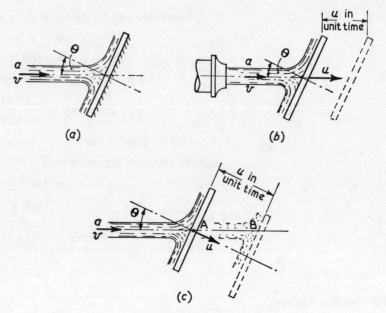

(a)

(b)

(c)

Figure 9.4

(b) The jet is continually extending so that only part of the flow leaving the nozzle strikes the plate (Fig. 9.4(b)).

$$\text{Increase in length of jet/sec} = u$$

$$\text{Mass required to extend jet/sec} = \rho a u$$

$$\text{Mass leaving nozzle/sec} = \rho a v$$

$$\text{Mass striking plate/sec} = \rho a (v - u)$$

$$\text{Initial component of velocity normal to plate} = v \cos \theta$$

$$\text{Final velocity normal to plate} = u \cos \theta$$

$$\text{Change of velocity normal to plate} = (v - u) \cos \theta$$

Force exerted normal to plate

$$= \text{mass striking/sec} \times \text{change of velocity}$$

$$= \rho a (v - u)^2 \cos \theta$$

Fig. 9.4(c) shows the arrangement for the numerical problem.

$$\text{Mass leaving nozzle per sec} = \rho a v$$

$$\text{Mass used to extend jet/sec} = \rho a \times (\text{AB}) = \rho a \times \frac{u}{\cos \theta}$$

$$\text{Mass striking plate/sec} = \rho a \left(v - \frac{u}{\cos \theta} \right)$$

$$\text{Initial component of velocity normal to plate} = v \cos \theta$$

$$\text{Final velocity normal to plate} = u$$

$$\text{Change of velocity normal to plate} = v \cos \theta - u$$

Force normal to plate = mass/sec × change of velocity

$$= \rho a \left(v - \frac{u}{\cos \theta} \right) (v \cos \theta - u)$$

$$P = \frac{\rho a (v \cos \theta - u)^2}{\cos \theta}$$

Putting $\rho = 1000 \, \text{kg/m}^3$, $a = \frac{1}{4}\pi(0\cdot025)^2 \text{m}^2$, $v = 6 \, \text{m/s}$, $u = 1\cdot5 \, \text{m/s}$, $\theta = 30°$

$$\text{Force normal to plate} = 1000 \times \tfrac{1}{4}\pi(0\cdot025)^2 \frac{(6 \cos 30° - 1\cdot5)^2}{\cos 30°}$$

$$= 7\cdot76\text{N}$$

Work done per sec = force × distance/sec

$$= Pu = 7\cdot76 \times 1\cdot5 = 11\cdot7\text{W}$$

K.E. per sec of jet at nozzle $= \frac{1}{2}(\rho a v^3)$

$$= (\tfrac{1}{2} \times 1000) \times \tfrac{1}{4}\pi(0\cdot025)^2 \times 6^3$$

$$= 53\cdot1\text{W}$$

$$\text{Efficiency} = \frac{\text{Work done/sec}}{\text{Energy of jet/sec}} = \frac{11\cdot7}{53\cdot1} = \textbf{22 per cent}$$

9.5 Hanging plate

A square plate of uniform thickness and length of side 30 cm hangs vertically from hinges at its top edge. When a horizontal jet strikes the plate at its centre, the plate is deflected and comes to rest at an angle of 30 deg to the vertical. The jet is 25 mm in diam and has velocity of 6 m/s. Calculate the mass of the plate and give the distance along the plate, from the hinge, of the point at which the jet strikes the plate in its deflected position.

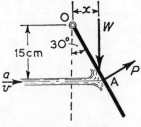

Figure 9.5

Solution. Fig. 9.5 shows the plate hinged at O in its deflected position.

Area of jet $= a = \frac{1}{4}\pi(0\cdot025)^2 \text{m}^2$, $v = 6\text{m/s}$

Mass striking plate per sec $= \rho a v$

$$= 1000 \times \tfrac{1}{4}\pi(0\cdot025)^2 \times 6$$

$$= 2\cdot94 \text{kg/s}$$

Change of velocity of jet normal to plate $= v \sin 60°$

$$= 6 \sin 60°$$

$$= 5·19\,\text{m/s}$$

Force normal to plate

$$= P$$

$$= \text{mass/sec} \times \text{change of velocity normal to plate}$$

$$= 2·94 \times 5·19 = \mathbf{15·25\,N}$$

Taking moment about O,

$$P \times \text{OA} = mgx$$

$$15·25 \times \frac{0·15}{\cos 30°} = m \times 9·81 \times 0·15 \tan 30°$$

$$m = \frac{15·25 \times 2}{9·81} = 3·11\,\text{kg}$$

$$\text{Distance OA} = 0·15 \times \frac{2}{\sqrt{3}} = 0·173\,\text{m} = \mathbf{17·3\,cm}$$

Force on curved vane

9.6 Curved vane

A jet of water having a velocity of 15 m/s impinges tangentially on to a stationary vane whose section is in the form of a circular arc with a subtended angle of 120°. Find the magnitude and direction of the reaction on the vane when the discharge from the jet is 0·45, assuming that the speed of the water relative to the vane remains constant. If now the vane is allowed to move with a velocity of 6 m/s in the direction of the jet, find the power developed for the same rate of discharge from the jet.

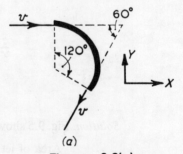

(a)

Figure 9.6(a)

Solution. The resultant force is found by determining its components on the X-direction and Y-direction (Fig. 9.6(*a*)) parallel and perpendicular to the incoming jet.

In the X-direction—

$$\text{Initial velocity} = v$$
$$\text{Final velocity} = -v \cos 60°$$
$$\text{Change of velocity in X-direction} = v(1 + \cos 60°) = 15(1 + \cos 60°)$$
$$\text{Mass of water deflected} = 0\cdot45 \, \text{kg/s}$$

Force in the X-direction

$$= F_x$$
$$= \text{mass/sec deflected} \times \text{change of velocity in X-direction}$$
$$= 0\cdot45 \times 15(1 + \cos 60°)$$
$$= 10\cdot1 \, \text{N}$$

In the Y-direction—

$$\text{Initial velocity of water} = 0$$
$$\text{Final velocity of water} = -v \sin 60°$$
$$\text{Change of velocity in Y-direction} = v \sin 60° = 15 \sin 60$$
$$\text{Mass of water deflected} = 0\cdot45 \, \text{kg/s}$$
$$\text{Force in the Y-direction} = F_y = 0\cdot45 \times 15 \sin 60° = 5\cdot84 \, \text{N}$$

From Fig. 9.6(b),

$$\text{Resultant force } P = \sqrt{(F_x^2 + F_y^2)}$$
$$= \sqrt{(10\cdot1^2 + 5\cdot84^2)} = \mathbf{11\cdot7 \, N}$$

If θ = inclination of P to incoming jet,

$$\tan \theta = \frac{F_y}{F_x} = \frac{5\cdot84}{10\cdot1} = 0\cdot578$$
$$\theta = \mathbf{30°}$$

When the vane moves with velocity $u = 6 \, \text{m/s}$ in direction of the incoming jet, the jet is extending by u metres each second.

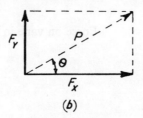

(b)

Figure   9.6(b)

If a = area of jet,

$$\text{Mass discharged from nozzle/sec} = \rho a v$$
$$\text{Mass required to extend jet/sec} = \rho a u$$

$\therefore$
$$\text{Mass deflected by vane/sec} = \rho a(v - u) = m$$
$$= \frac{v - u}{v} \times \text{mass discharged from nozzle/sec}$$
$$= \frac{15 - 6}{15} \times 0\cdot45 = 0\cdot27 \, \text{kg/s}$$

Initial absolute velocity in direction of incoming jet $= v$. At the entry to the vane the water will have a relative velocity $v_r = v - u$ relative to the vane, which is stated to be constant so that at outlet

Velocity relative to the vane $= v_r = v - u$

Component of v_r in X-direction $= -(v - u) \cos 60°$

Velocity of vane in X-direction $= u$

Component of absolute velocity at outlet in X-direction

$$= u - (v - u \cos 60°)$$

The triangles of velocities corresponding to the above are shown in Fig. 9.6(c).

Change of absolute velocity in X-direction

$$= v - u + (v - u) \cos 60°$$

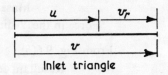

Inlet triangle

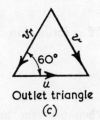

Outlet triangle

(c)

Figure 9.6(c)

Force on vane in direction of motion $= F_x$

$$= m(v - u)(1 + \cos 60°)$$

$$= 0{\cdot}27(15 - 6) \times 1{\cdot}5$$

$$= 3{\cdot}65\,\text{N}$$

Work done per second

$$= \text{Force in direction of motion} \times \text{velocity of vane}$$

$$= F_x \times u$$

$$= 3{\cdot}65 \times 6 = 21{\cdot}9\,\text{W}$$

9.7 Curved vane considering friction

A 15 mm diam nozzle ($C_v = 0.97$) is supplied with water under a head of 30 m. The jet acts on a fixed vane, the water sliding on to the plate tangentially and being turned through 165 deg. Calculate the force on the vane in the direction of the jet (*a*) if there is no friction, (*b*) if the velocity of the water leaving the vane is 0.8 of its impinging velocity.

Solution. The jet is deflected as shown in Fig. 9.7.

$$\text{Initial velocity } V_1 = C_v\sqrt{(2gh)} \quad \text{where } h = \text{head at nozzle}$$

$$= 0.97\sqrt{(2g \times 30)} = 23.5\,\text{m/s}$$

$$\text{Final velocity in the direction of the jet} = V_2 \cos 165°$$

$$\text{Change of velocity in the direction of the jet} = V_1 - V_2 \cos 165°$$

$$= V_1 + V_2 \cos 15°$$

$$\text{Mass deflected by vane/sec} = \rho a V_1$$

where $a = $ jet area

$$M = 1000 \times \tfrac{1}{4}\pi(0.015)^2 \times 23.5 = 4.14\,\text{kg/s}$$

Force in direction of jet

$$= \text{mass deflected/sec} \times \text{change of velocity in direction of jet}$$

$$= M(V_1 + V_2 \cos 15°)$$

Figure 9.7

(*a*) If there is no friction

$$V_1 = V_2 = 23.5\,\text{m/s}$$

$$\text{Force in direction of jet} = 4.14 \times 23.5(1 + \cos 15°)$$

$$= \mathbf{191\,N}$$

(*b*) If $V_2 = 0.8V_1$

$$\text{Force in direction of jet} = 4.14 \times 23.5(1 + 0.8 \cos 15°)$$

$$= \mathbf{173\,N}$$

A jet of water 75 mm in diam with a velocity of 12 m/s meets a vane having a velocity of 4·8 m/s in the direction of the jet. If the water meets the vane tangentially and is deflected through 120 deg, find (*a*) the force in the direction of the jet, (*b*) the work done per second on a single vane moving in a straight line, (*c*) the work done per second on a series of vanes mounted on the circumference of a large wheel.

Solution. Fig. 9.8 shows the deflection of the jet and the triangle of relative velocities at inlet and outlet.

$$V_1 = \text{absolute velocity at inlet}$$

$$u = \text{velocity of vane}$$

$$v_r = \text{velocity of water relative to vane}$$

From the inlet triangle $\quad v_r = V_1 - u$

$$V_2 = \text{absolute velocity at exit}$$

If there is no friction, relative velocity at exit = relative velocity at entry = v_r.

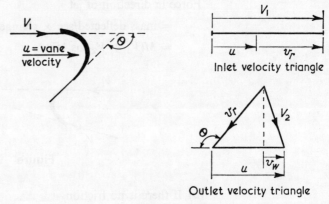

Inlet velocity triangle

Outlet velocity triangle

Figure 9.8

To find the force in the direction of the jet, it is necessary to find the component of the absolute velocity V_2 in the direction of the jet which is v_w.

From the outlet triangle

$$v_w = u - v_r \cos (180 - \theta)$$

Change of absolute velocity in direction of jet

$$= V_1 - v_w$$
$$= V_1 - u + v_r \cos (180 - \theta)$$
$$= V_1 - u + (V_1 - u) \cos 60°$$
$$= (V_1 - u)(1 + \cos 60°)$$

(a) For a single moving vane the jet is continually increasing in length at the rate of u m/s.

$$\text{Mass deflected by vane/sec} = \rho a (V_1 - u)$$

$\therefore$ Force in direction of jet

$$= \text{mass deflected/sec} \times \text{change of velocity}$$

$$= \rho a (V_1 - u)^2 (1 + \cos 60°)$$

Putting $\rho = 1000 \, \text{kg/m}^3$, $a = \frac{1}{4}\pi(0.075)^2 \, \text{m}^2$, $V_1 = 12 \, \text{m/s}$, $u = 4.8 \, \text{m}$

Force in direction of jet

$$= P$$

$$= 1000 \times \tfrac{1}{4}\pi(0.075)^2 \times (12 - 4.8)^2 \times (1 + \tfrac{1}{2})$$

$$= \tfrac{1}{4}\pi \times 5.63 \times 52 \times 1.5 = \textbf{343N}$$

(b) Work done per second on a single vane

$$= \text{force} \times \text{velocity of vane}$$

$$= P \times u = 343 \times 4.8 = \textbf{1645W}$$

(c) For a series of vanes the length of the jet remains constant.

$$\text{Mass deflected by vane per second} = \rho a V_1$$

$$\text{Force in direction of jet} = \rho a V_1 (V_1 - u)(1 + \cos 60°)$$

Work done per second

$$= \rho a V_1 (V_1 - u)(1 + \cos 60°)u$$

$$= 1000 \times \tfrac{1}{4}\pi \times (0.675)^2 \times 12 \times 7.2 \times 1.5 \times 4.8$$

$$= \textbf{2740W}$$

9.9 Inlet angle for no shock

A jet of water discharging 22.5 kg/s with velocity 21 m/s impinges on a series of curved vanes moving at 12 m/s in a direction making 25 deg to that of the jet. Determine the angles of the leading edges of the vane so that the water enters without shock.

If the water leaves the vanes at an angle of 150 deg to the direction of motion (*when the vane is stationary*), calculate the force in this direction, assuming that all the water is effective and that its velocity relative to the blade is reduced by 20 per cent during its passage of the blade. Make a sketch of the arrangement.

Solution. If the water enters without shock the relative velocity at inlet v_{1r} must be tangential to the blade at an angle α to the direction of motion (Fig. 9.9).

$$v_1 = \text{absolute velocity of incoming jet}$$

$$u = \text{velocity of the vanes}$$

$$v_{1r} = \text{relative velocity of jet over the blade at inlet}$$

$$v_{2r} = \text{relative velocity of water at outlet}$$

$$v_2 = \text{absolute velocity of water at outlet}$$

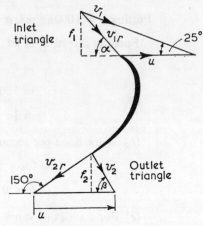

Figure   9.9   ($u = 12\,\text{m/s}$)

From the inlet triangle,

$$f_1 = 21 \sin 25° = 8 \cdot 9\,\text{m/s}$$

$$\tan \alpha = \frac{f_1}{21 \cos 25° - 12} = \frac{8 \cdot 9}{7 \cdot 0} = 1 \cdot 265$$

Blade angle at inlet $= \alpha = \mathbf{51° \ 40'}$

Force in direction of motion

$$= \text{mass/sec} \times \text{change of velocity in direction of motion}$$

Mass deflected per second $= 22 \cdot 5\,\text{kg/s}$

since there is a series of vanes and the average length of the jet is therefore constant.

Change of velocity in direction of motion $= v_1 \cos 25° - v_2 \cos \beta$

To find v_2 and β consider the outlet triangle:

$$v_{2r} = 0 \cdot 8 v_{1r} = 0 \cdot 8\,\frac{f_1}{\sin \alpha} = \frac{0 \cdot 8 \times 8 \cdot 9}{\sin 51° \ 40'} = 9 \cdot 06\,\text{m/s}$$

$$f_2 = v_{2r} \sin 30° = 9 \cdot 06 \sin 30° = 4 \cdot 53\,\text{m/s}$$

$$\tan \beta = \frac{f_2}{u - v_{2r} \cos 30°} = \frac{4 \cdot 53}{12 - 9 \cdot 06 \cos 30°} = \frac{4 \cdot 53}{4 \cdot 18} = 1 \cdot 086$$

$$\beta = 47° \ 22'$$

$$v_2 = \frac{f_2}{\sin \beta} = \frac{4.53}{\sin 47° 22'} = 6.15 \, \text{m/s}$$

Force in direction of motion

$$= 22.5(21 \cos 25° - 6.15 \cos 47° 22')$$
$$= 22.5(19 - 4.2) = \textbf{333N}$$

Forces on reducers and bends

9.10 Force on reducing bend

> Derive an expression for the magnitude and direction of the resultant force on a horizontal reducing bend which deflects the flow of a fluid of specific weight w through an angle θ, if at entry the pressure is p_1, area A_1, and velocity v_1 and at exit the corresponding values are p_2, A_2 and v_2, the discharge being Q.
>
> A 600 mm main carries water under a head of 30 m with a velocity of flow of 3 m/s. The main is fitted with a bend which turns the axis through 75 deg. Calculate the resultant force on the bend.

Solution. Fig. 9.10(*a*) shows the reducing bend. The resultant force P on the bend can be found by combining the components P_X and P_Y as shown in Fig. 9.10(*b*).

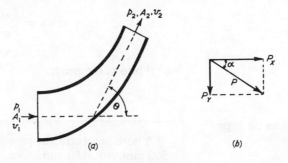

Figure 9.10

The forces acting in the direction of P_X are

Force due to p_1 in direction of $P_X = p_1 A_1$

Component of force due to p_2 in direction of $P_X = -p_2 A_2 \cos \theta$

Force due to change of momentum in direction of P_X

$$= \text{mass/sec} \times \text{change of velocity in direction of } P_X$$
$$= \rho Q(v_1 - v_2 \cos \theta)$$

$$\therefore \qquad P_X = p_1 A_1 - p_2 A_2 \cos \theta + \rho Q(v_1 - v_2 \cos \theta)$$

The forces acting in the direction of P_Y are

$$\text{Force due to } p_2 \text{ in direction of } P_Y = p_2 A_2 \sin \theta$$

$$\text{Force due to change of momentum in direction of } P_Y = \rho Q v_2 \sin \theta$$

$$P_Y = p_2 A_2 \sin \theta + \rho Q v_2 \sin \theta$$

$$\text{Resultant force } P = \sqrt{(P_X{}^2 + P_Y{}^2)}$$

and

$$\tan \alpha = \frac{P_Y}{P_X}$$

Putting $p = p_1 = p_2 = \rho g h = 1000 \times 9.81 \times 30 \,\mathrm{N/m^2}$, $a = a_1 = a_2 = \frac{1}{4}\pi \times 0.6^2 = 0.282\,\mathrm{m^2}$, $v = v_1 = v_2 = 3\,\mathrm{m/s}$, $Q = av$, $\theta = 75°$

$$P_X = (ap + \rho a v^2)(1 - \cos \theta)$$

$$= a(p + \rho v^2)(1 - \cos \theta)$$

$$= 0.282(1000 \times 9.81 \times 30 + 1000 \times 9)(1 - \cos 75°)$$

$$= 0.282 \times 304000 \times 0.741 = 63400\,\mathrm{N}$$

$$P_Y = (pa + \rho a v^2) \sin \theta = a(p + \rho v^2) \sin \theta = \rho a(gh + v^2) \sin \theta$$

$$= 1000 \times 0.282(9.81 \times 30 + 9)\,0.966\,\mathrm{N}$$

$$= 82800\,\mathrm{N}$$

$$\text{Resultant force } P \text{ on bend} = \sqrt{(P_X{}^2 + P_Y{}^2)}$$

$$= \sqrt{(3990 \times 10^6 + 6856 \times 10^6)}$$

$$= 104160\,\mathrm{N}$$

$$\tan \alpha = \frac{P_Y}{P_X} = \frac{82800}{63400} = 1.30$$

$$\alpha = 52° \, 30'$$

and P bisects the bend.

9.11 Force on taper pipe

A horizontal straight pipe gradually reduces in diameter from 300 mm to 150 mm. Neglecting friction, find the total longitudinal thrust on the pipe if at the larger end the pressure is 275 kN/m² and the velocity is 3 m/s. The pipe is conveying water.

Solution. Let p_1, v_1, and A_1 be the pressure, velocity and area at the larger end and p_2, v_2 and A_2 the corresponding values at the smaller end.

For continuity of flow

$$A_1 v_1 = A_2 v_2$$

$$\tfrac{1}{4}\pi \times (0.300)^2 \times 3 = \tfrac{1}{4}\pi \times (0.150)^2 \times v_2$$

$$v_2 = 12\,\mathrm{m/s}$$

Neglecting frictional losses, by Bernoulli's theorem

$$\frac{p_1}{w} + \frac{v_1{}^2}{2g} = \frac{p_2}{w} + \frac{v_2{}^2}{2g}, \qquad p_2 = p_1 + \tfrac{1}{2}\rho(v_1{}^2 - v_2{}^2)$$

$$p_2 = 275 \times 10^3 + \tfrac{1}{2} \times 1000(3^2 - 12^2)$$
$$= 275 \times 10^3 - 68 \times 10^3 = 207 \times 10^3\,\text{N/m}^2$$

From Example 9.10, when $\theta = 0$

$$P = P_x = p_1 A_1 - p_2 A_2 = \rho A_1 v_1 (v_2 - v_1)$$

Putting $p_1 = 275 \times 10^3\,\text{N/m}^2$, $p_2 = 207 \times 10^3\,\text{N/m}^2$, $A_1 = \tfrac{1}{4}\pi \times (0\cdot300)^2$ m^2, $A_2 = \tfrac{1}{4}\pi(0\cdot150)^2$m^2, $v_1 = 3$m/s, $v_2 = 12$m/s

Force in direction of motion

$$= 275 \times 10^3 \times \tfrac{1}{4}\pi(0\cdot3)^2 - 207 \times 10^3 \times \tfrac{1}{4}\pi(0\cdot15)^2$$
$$- 1000 \times \tfrac{1}{4}\pi(0\cdot3)^2 \times 3 \times 9$$
$$= 19\cdot45 \times 10^3 - 3\cdot65 \times 10^3 - 1\cdot9 \times 10^3$$
$$= \mathbf{13\cdot9 \times 10N}$$

Reaction of a jet

9.12 Rotating nozzles

> Water is discharged tangentially from two nozzles at opposite ends of an arm which has a length L of $0\cdot6$ m and is pivoted at its centre. The velocity of discharge relative to the nozzle v is 6 m/s and the diameter d of each nozzle is $12\cdot5$ mm.
>
> (a) Calculate the torque exerted when the arm is stationary.
>
> (b) Find an expression for the work done per second and the efficiency of the device when the arm is allowed to rotate with a peripheral velocity u.
>
> (c) If the velocity of discharge relative to the nozzle remains 6 m/s, what must be the value of u for maximum power and what is the work done per second and the efficiency under these conditions?

Solution. Fig. 9.11 shows the arrangement of the nozzles.

(a) Force exerted by reaction of jet = rate of change of momentum
$$= \text{mass/sec} \times \text{velocity given to jet}$$
$$= \rho a v^2 \quad \text{where} \quad a = \text{area of jet.}$$

Equal and opposite forces are exerted at each nozzle.

$$\text{Turning moment} = \rho a v^2 L$$

$\rho = 1000\,\text{kg/m}^3$, $a = \tfrac{1}{4}\pi d^2 = \tfrac{1}{4}\pi(0\cdot0125)^2$ m^2, $v = 6$m/s, $L = 0\cdot6$m

$\therefore \qquad \text{Turning moment} = 1000 \times \tfrac{1}{4}\pi(0\cdot0125)^2 \times 36 \times 0\cdot6$
$$= \mathbf{2\cdot65\,\text{N-m}}$$

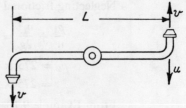

Figure 9.11

(*b*) When the arm has a peripheral velocity u

$$\text{Absolute velocity given to jet} = v - u$$

$$\text{Mass discharged per second for one nozzle} = \rho a v$$

$$\text{Force due to reaction of jet} = \rho a v (v - u)$$

Work done per second by one nozzle

$$= \text{force} \times \text{velocity of nozzle}$$

$$= \rho a v (v - u) u$$

$$\text{Total work done per second for two nozzles} = 2\rho a v (v - u) u$$

$$\text{Absolute velocity of issuing jets} = v - u$$

$$\text{Kinetic energy/sec lost in jets} = 2 \times \tfrac{1}{2}\rho a v (v - u)^2$$

Total energy supplied per second

$$= \text{Work done per sec} + \text{K.E. lost}$$

$$= 2\rho a v \{(v - u)u + \tfrac{1}{2}(v - u)^2\}$$

$$= 2\rho a v \{\tfrac{1}{2}(v^2 - u^2)\}$$

$$\text{Efficiency} = \frac{\text{Work done/sec}}{\text{Energy supplied/sec}} = \frac{2\rho a v (v - u)u}{2\rho a v \{\tfrac{1}{2}(v^2 - u^2)\}} = \frac{2u}{v + u}$$

(*c*) $$\text{Power} = 2\rho a v (v - u) u$$

For a given value of v the power will be a maximum for the value of u which makes $(v - u)u$ a maximum. Differentiating and equating to zero

$$\frac{d}{du}(v - u)u = v - 2u = 0$$

For maximum power $\qquad u = \tfrac{1}{2}v = 3\,\text{m/s}$

$$\text{Work done/sec} = 2\rho a v (v - u)(u)$$

$$= 2 \times 1000 \times \tfrac{1}{4}\pi (0 \cdot 0125)^2 \times 6 \times 3 \times 3$$

$$= \mathbf{13 \cdot 25\,J/s}$$

$$\text{Efficiency} = \frac{2u}{v + u} = \frac{2u}{3u} = \mathbf{66\tfrac{2}{3}\ \textbf{per cent}}$$

9.13 Jet propulsion

Obtain an expression for the efficiency of a jet-propelled vessel in which the water is drawn in through the bows and expelled at the stern, in terms of u the velocity of the vessel, and v the velocity of the jet relative to the vessel.

The resistance to motion of a vessel is 22 000 N at a velocity of $4 \cdot 5$ m/s. The jet efficiency is to be 80 per cent and the mechanical efficiency of the pumps is 75 per cent. Hydraulic losses in the ducts are 5 per cent of the relative kinetic energy at exit. Determine (a) the velocity of the jet, (b) the orifice area at exit, (c) the power required to drive the pumps for the given speed of the vessel.

Solution.

$$\text{Absolute incoming velocity of water} = 0$$

$$\text{Absolute outgoing velocity of water} = v - u$$

$$\text{Change of velocity} = v - u$$

$$\text{Mass of fluid expelled per second} = \rho a v$$

$$\text{Propelling force} = \text{rate of change of momentum}$$

$$= \rho a v (v - u)$$

$$\text{Work done per second by jet} = \rho a v (v - u) u$$

$$\text{Kinetic energy of water at inlet per sec} = \tfrac{1}{2} \rho a v u^2$$

$$\text{Kinetic energy of water at outlet per sec} = \tfrac{1}{2} \rho a v v^2$$

$$\text{Energy supplied per second} = \tfrac{1}{2} \rho a v (v^2 - u^2)$$

$$\text{Efficiency} = \frac{\text{Work done per sec}}{\text{Energy supplied per sec}} = \frac{(v - u)u}{\tfrac{1}{2}(v^2 - u^2)} = \frac{2u}{v + u}$$

(a) If $u = 4 \cdot 5$ m/s and the efficiency $= 80$ per cent $= 0 \cdot 8$

$$0 \cdot 8 = \frac{2u}{v + u} = \frac{2 \times 4 \cdot 5}{v + 4 \cdot 5},$$

$$v = 6 \cdot 75 \text{m/s}$$

(b) Work done per sec $=$ resistance $\times$ velocity of vessel

$$= 22000 \times 4 \cdot 5 = 99000 \, \text{W}$$

$$= \rho a v (v - u) u$$

$$\therefore \qquad a = \frac{99000}{\rho v (v - u) u}$$

$$= \frac{99000}{1000 \times 6 \cdot 75 \times 2 \cdot 25 \times 4 \cdot 5}$$

$$\text{Area of orifice at exit} = a = \mathbf{1 \cdot 45 m^2}$$

(c) $\qquad$ Power supplied to jet $= \tfrac{1}{2} \rho a v (v^2 - u^2)$

$$\text{Loss of energy in ducts} = 0.05 \frac{v^2}{2g}$$

Total power required from pumps

$$= \tfrac{1}{2}\rho a v\{(v^2 - u^2) + 0.05v^2\}$$

$$= \tfrac{1}{2} \times 1000 \times 1.45 \times 6.75\{(6.75^2 - 4.5^2) + 0.05 \times 6.75^2\}$$

$$= 1000 \times 4.9(25.25 + 2.27) = 134.5 \times 10^3\,\text{W}$$

If the mechanical efficiency of the pumps is 75 per cent

$$\text{Power required to drive pumps} = \frac{100}{75} \times 134.5 \times 10^3\,\text{W}$$

$$= \mathbf{179.5\,kW}$$

9.14 Jet propulsion

A vessel is propelled by the reaction of jets discharged astern, the water being drawn in initially at the side. Establish expressions for the theoretical efficiency and for the input of power to the pumps in terms of the speed of the ship s, the velocity through the jet v relative to the ship, the weight of water pumped per second W, and the combined efficiency of the pump and pipe system η.

In an actual case a small ship is fitted with jets of total area 0.65 m². The velocity through the jets is 9 m/s and the ship's speed is 18.5 km/h. The engine efficiency is 85 per cent, the pump efficiency is 65 per cent and the pipe losses are equal to 10 per cent of the kinetic energy of the jets. Determine the propelling force and the overall efficiency. Density of sea water = 1025 kg/m³.

Solution.

$$\text{Mass of fluid expelled per second} = \frac{W}{g}$$

$$\text{Absolute velocity of incoming water} = 0$$

$$\text{Absolute velocity of outgoing water} = v - s$$

$$\text{Change of velocity} = v - s$$

$$\text{Propulsive force} = \frac{W}{g}(v - s)$$

$$\text{Work done by jets/sec} = \frac{W}{g}(v - s)s$$

When the intake is at the side the water enters without kinetic energy and leaves with velocity v.

$$\text{Energy supplied by pumps} = \frac{Wv^2}{2g}$$

$$\text{Theoretical efficiency} = \frac{W/g(v-s)s}{(W/2g)v^2} = \frac{2s(v-s)}{v^2}$$

$$\text{Power delivered by pumps} = \frac{Wv^2}{2g}$$

$$\text{Power input to pumps} = \frac{Wv^2}{2g \times \eta}$$

$$\text{Jet speed } v = 9\,\text{m/s}$$

$$\text{Ship speed } s = 18 \cdot 5\,\text{km/h} = 5 \cdot 16\,\text{m/s}$$

$$\text{Mass/sec discharged} = \frac{W}{g} = \rho a v = 1025 \times 0 \cdot 65 \times 9 = 6000\,\text{kg/s}$$

$$\text{Propelling force} = \frac{W}{g}(v-s) = 6000 \times 3 \cdot 84$$

$$= 23040\,\text{N}$$

If 10 per cent of the kinetic energy of jet is lost in pipes

$$\text{Energy supplied to jets per second} = 1 \cdot 1 \frac{Wv^2}{2g}$$

$$\text{Efficiency of jet and pipes} = \frac{2s(v-s)}{1 \cdot 1v^2}$$

$$= \frac{2 \times 5 \cdot 16 \times 3 \cdot 84}{1 \cdot 1 \times 81}$$

$$= 44 \cdot 5 \text{ per cent}$$

$$\text{Overall efficiency} = 0 \cdot 85 \times 0 \cdot 65 \times 0 \cdot 445$$

$$= 24 \cdot 6 \text{ per cent}$$

Problems

1 An undershot waterwheel consists of a series of flat vanes, mounted radially on a wheel of large diameter, which are struck normally by a jet of water $0 \cdot 3$ m in diam. If the velocity of the water leaving the nozzle is $7 \cdot 5$ m/s and the velocity of the vanes is $4 \cdot 8$ m/s what is the force exerted by the jet on the vanes, the work done per second and the efficiency?
Answer 1433 N; $6 \cdot 86$ kW; 46 per cent

2 A flat plate is struck normally by a jet of water 50 mm in diam with a velocity of 18 m/s. Calculate (*a*) the force on the plate when it is stationary, (*b*) the force on the plate when it moves in

the same direction as the jet as 6 m/s, (c) the work done per second and the efficiency in case (b).
Answer 636 N; 283 N; 1698 W; 29·6 per cent

3 A square plate, mass 12·7 kg, of uniform thickness and 300 mm edge, is hung so that it can swing freely about its upper horizontal edge. A horizontal jet 20 mm in diam strikes the plate with a velocity of 15 m/s and the centreline of the jet is 150 mm below the upper edge of the plate so that when the plate is vertical the jet strikes the plate normally at its centre.

Find (a) what force must be applied at the lower edge of the plate to keep it vertical, (b) what inclination to the vertical the plate will assume under the action of the jet if allowed to swing freely.
Answer 35·34 N, 34·57 deg

4 A rectangular plate of mass 5·45 kg is suspended vertically by a hinge on the top horizontal edge. The centre of gravity of the plate is 10 cm from the hinge. A horizontal jet of water of 25 mm diam, whose axis is 15 cm below the hinge, impinges normally on the plate with a velocity of 5·65 m/s. Find the horizontal force applied at the centre of gravity to maintain the plate in its vertical position. Find the alteration of the velocity of the jet if the plate is deflected through an angle of 30 deg, and the same horizontal force continues to act at the centre of gravity of the plate.
Answer 23·5 N; 2·31 m/s increase

5 A jet of water, 50 mm diam, with a velocity of 18 m/s, strikes a flat plate inclined at an angle of 25 deg to the axis of the jet. Determine the normal force exerted on the plate (a) when the plate is stationary, (b) when the plate is moving at 4·5 m/s in the direction of the jet, and (c) determine work done per second and the efficiency for case (b).
Answer 269 N; 151·5 N; 288 W; 5 per cent

6 The flow through an experimental fan is discharged along a cylindrical duct from which it passes into the atmosphere through an opening controlled by a cone. The cone is mounted co-axially with the cylinder and the opening may be varied by varying the axial position of the cone.

The figure shows the form of the flow at a particular setting of the cone, it being assumed that there is no mixing between the atmosphere and the annular jet discharging over the conical surface. The pressure in the cylindrical duct a short distance upstream of the opening is 113 kN/m² and the atmospheric pressure is 106 kN/m². Calculate the axial force exerted on the cone. The effects of compressibility and viscosity may be neglected, and the air density assumed constant at 1·19 kg/m³.
Answer 7·18 kN

7 A jet of water, 75 mm diam, strikes a flat plate with a velocity of 24 m/s. The normal to the plate is inclined at 30 deg to the axis of the jet. Calculate the normal force on the plate (a) when the

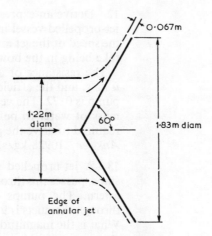

Figure 9.12

plate is stationary, (*b*) when the plate has a velocity of 12 m/s in the same direction as the jet.
Answer 2·2 kN: 0·55 kN

8 The nozzle of a fire-hose produces a 50 mm diam jet of water. If the discharge is 85 dm³/s, calculate the reaction of the jet if the jet velocity is 10 times the velocity of the water in the hose.
Answer 3320 N

9 A tanker discharges a jet of water horizontally backwards with a velocity of 4·8 m/s. If the rate of discharge is 85 dm³/s, what force is required to keep the tanker at rest?
  If the tanker is allowed to move forward with a uniform velocity of 1·8 m/s, and the jet velocity relative to the tanker is still 4·8 m/s, what will be the force on the tanker?
Answer 407 N; 407 N

10 An experimental orifice, 40 mm in diam, is mounted in the side of a tank, which is supported in knife edges at points 1·5 m above the orifice. A lever with a sliding weight is fixed to the tank on the side opposite to the orifice. For a discharge of 6·5 kg/s under a head of 1·8 m the sliding weight of 18 kg had to be moved away from the fulcrum a distance of 370 mm from its zero position. Calculate the coefficients of the orifice.
Answer 0·99; 0·870; 0·879

11 Derive a formula for the propulsion efficiency of a jet-propelled boat in which the water enters through tubes facing the direction of motion and is expelled through an orifice directed dead astern. A boat fitted with this arrangement requires 15 kW to propel it through the water at 9 m/s. If the propulsion efficiency is to be 80 per cent, calculate the velocity of the jet, the area of inlet and exit pipes and the quantity of water passing through the pumps per second.
Answer 4·5 m/s; 0·0329 m²; 0·0219 m²; 0·269 m³/s

12 Derive an expression for the efficiency of the jet drive of a jet-propelled vessel in terms of u, the speed of the vessel, and v the speed of the jet measured relatively to that of the ship, the intake being in the bows pointing forward.

The resistance of a ship is given by $5 \cdot 55 \, u^6 + 978 \, u^{1 \cdot 9}$ N at u m/s, and the efficiency of the jet drive is $0 \cdot 8$ while that of the pump is $0 \cdot 72$. The vessel is to be driven at $3 \cdot 4$ m/s. Find (*a*) the mass of water to be pumped astern per second, (*b*) the power required to drive the pump.
Answer 10928 kg/s; 109·7 kW

13 A jet-propelled vessel takes in water through ducts amidships, passes this through pumps, and discharges it through ducts astern. The pumps deliver 34 m³/min; the velocity of flow through the ducts is 9 m/s and the speed of the vessel is $4 \cdot 5$ m/s. What is the magnitude of the propulsive force?
Answer 2550 N

14 A jet of water, $6 \cdot 5$ cm² in cross-sectional area, moving at 12 m/s, is turned through an angle of 135 deg by a curved plate. The plate is moving at $4 \cdot 5$ m/s in the same direction as the jet. Neglecting any loss of velocity by shock or friction, find the amount of work done on the plate per sec.
Answer 281 W

15 A jet of water 75 mm in diam with velocity 21 m/s flows tangentially onto a stationary vane which deflects it through 120 deg. What is the magnitude and direction, referred to the direction of the jet, of the resultant force on the vane?

If this jet flows onto a series of vanes similarly oriented with regard to it but moving in the direction of the jet with a velocity of $10 \cdot 5$ m/s, determine (*a*) the force on the system of vanes in the direction of motion, (*b*) the work done per sec, (*c*) the efficiency.
Answer 3375 N, 30 deg; 1460 N, 15 350 W, 75 per cent

16 A jet of water of 100 mm diam is moving at 36 m/s and is deflected by a vane moving at 15 m/s in a direction at 30 deg to the direction of the jet. The water leaves the vane with no velocity in the direction of motion of the vane. Determine (*a*) the inlet and outlet angles of the vane for no shock at entry or exit, (*b*) the force on the vane in the direction of motion, and (*c*) the force on the vane at right angles to the direction of motion. Take the outlet velocity of the water relative to the vane to be $0 \cdot 85$ of the relative velocity at entry.
Answer 48° 3′; 43° 11′; 5·63 kN; 0·7 kN

17 A jet of water of 50 mm diam, having a velocity of 24 m/s impinges tangentially on a series of vanes which when stationary deflect the jet through an angle of 120 deg.

Calculate the magnitude of the force on the vanes in the direction of motion when they are (*a*) stationary, (*b*) moving with a velocity of 9 m/s in the same direction as the jet. In case (*b*) determine also the work done per second in the vanes and the efficiency.
Answer 1696 N; 1062 N, 9550 W, 70·2 per cent

18 A nozzle, having a coefficient of velocity of 0·96, operates under a head of 90 cm of water and directs a 50 mm diam jet of water on to a ring of axial flow impulse blades, which have an inlet angle of 40 deg measured relative to the direction of blade motion and which turn the water through an angle of 105 deg during its passage over the blades (i.e. blade angle at outlet = 35 deg). Because of friction the velocity of the water relative to the blades at outlet is only 0·85 of that at inlet. The blade speed is to be 18 m/s and the water is to flow on to the blades without shock.

Draw the velocity diagram to a scale of 1 cm = 2·5 m/s per sec and determine —

(a) The angle which the line of the jet should make with the direction of motion of the blades.
(b) The power developed by the blade ring.
Answer 22° 18′; 51·1 kW

19 A jet of water delivers 85 dm³/s at 36 m/s on to a series of vanes moving in the same direction at 18 m/s. If stationary the vanes would divert the jet through 135 deg. Friction reduces the relative velocity at exit to 0·80 of that at entrance. Determine the magnitude of the resultant force on the vane and the efficiency of the arrangement. Assume that the water enters without shock.
Answer 2546 N: 0·783

20 A jet of water, flowing at the rate of 20 kg/s and at 25 m/s velocity, impinges on a series of vanes moving at 12 m/s in a direction making an angle of 25° to the jet. Determine the inlet angle of the blade for no shock entry.

If the vane exit angle is 150° to the direction of motion, calculate the force in the direction of motion and the power developed if friction reduces the water velocity relative to the vanes by 20 per cent during its passage over the vanes.
Answer 45 deg, 381·2 N, 4570 W

21 A jet of water leaves a 20 mm diam nozzle at 36 m/s and enters a series of vanes without shock. The vanes move in the same direction as the jet at 15 m/s and are so shaped that they would, if stationary, deflect the jet through an angle of 150 deg. Frictional resistance reduces the velocity of the water relative to the vanes by 12 per cent as the water flows over the vanes. Working from first principles, calculate (a) the magnitude and direction of the resultant force on the vanes, (b) the power developed by the arrangement.
Answer 431 N at 14° to jet direction; 6·27 kW

22 A jet of water discharges 13·6 kg/s at 24 m/s in a direction making 30 deg to the direction of a series of curved vanes moving at 10·5 m/s. If the outlet angle of the vanes is 20 deg, determine (a) the inlet angle of the vanes such that there is no shock at entry, (b) the work done per second.
Answer 49° 25′; 3·59 kW

23 A 1 m diam pipe is deflected through 90 deg, the ends being anchored by rods at right angles to the pipe at the ends of the bend. If the pipe is delivering $1 \cdot 78$ m³/s, find the tension in each tie rod.

Answer   4010 N

24   Water flows through a $0 \cdot 9$ m diam pipe at the end of which there is a reducer connecting to a $0 \cdot 6$ m diam pipe. If the gauge pressure at the entrance to the reducer is 414 kN/m² and the velocity is $2 \cdot 1$ m/s, determine the resultant thrust on the reducer, assuming that the frictional loss of head in the reducer is $1 \cdot 5$ m.

Answer $149 \cdot 5$ kN

25   A bend in a pipeline gradually reduces from 600 mm to 300 mm diam and deflects the flow of water through an angle of 60 deg. At the larger end the gauge pressure is 172 kN/m². Determine the magnitude and direction of the force exerted on the bend, (*a*) when there is no flow, (*b*) when the flow is $0 \cdot 85$ m³/s.

Answer $43 \cdot 8$ kN; 13° 53'; 45 kN; 19° 46'

26   The cascade of vanes shown is used at the corner of a wind tunnel to deflect the air through a right angle and also to accelerate it from $13 \cdot 71$ m/s to $18 \cdot 3$ m/s. The depth of the rectangular duct measured parallel to the span of the cascade is $1 \cdot 83$ m throughout, and the breadth is $1 \cdot 83$ m upstream and $1 \cdot 371$ m downstream of the cascade. The spacing of the vanes is $0 \cdot 254$ m measured along the cascade.

Calculate the magnitude and direction of the force on each vane of the cascade. Neglect the effects of viscosity and assume that the density of air is $1 \cdot 29$ kg/m³.

Answer   152 N at $36 \cdot 8°$ to approaching stream

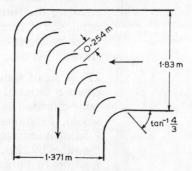

Figure   9.13

27   Fig. 9.14 shows part of a horizontal pipeline $0 \cdot 3$ m in diam, the bend being so short that it may be considered as included in the 54 m length. The head at (1) is 36 m and the frictional coefficient $f = 0 \cdot 005$ and the speed of the water in the pipe is $3 \cdot 6$ m/s. Both joints (1) and (2) are free to expand axially. Find (*a*) the force on the pipe in the direction of the 54 m length, (*b*) the force at right angles to this direction.

Answer $8 \cdot 77$ kN; $17 \cdot 15$ kN

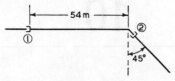

Figure   9.14

28   Fig. 9.15 shows a vertical main A carrying an unsupported branch D which is 2·4 m long and which is connected through a bend B to another pipe C, the connecting joint between B and C allowing pipe C to expand freely. The pressure in A is 275 kN/m² above atmospheric and the rate of flow in pipe D which has a bore of 150 mm is 2·4 m/s. Neglecting friction, find the torque set up on the vertical main and the force on A in the direction of the length of pipe D, (*a*) if the bend B is of 150 mm bore throughout, (*b*) if the bend tapers from 150 mm to 100 mm in diam at its discharge end, the diameter of pipe C being correspondingly reduced.

Answer   (*a*) 2468 Nm, 1454 N; (*b*) 2057 Nm, 3340 N

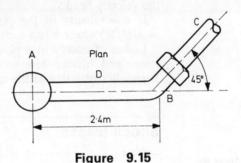

Figure   9.15

10

Losses of energy in pipelines

A pipe is defined as a closed conduit of circular section through which the fluid flows filling the complete cross-section. The fluid in the pipe has no free surface. It will be at a pressure above or below atmospheric and this pressure may vary along the pipe.

Losses of energy in a pipe line are due to

(a) Shock from the disturbance of the normal flow due to bends or sudden changes of section, and

(b) Frictional resistance to flow.

These losses are conveniently expressed as energy lost in N-m/N, that is to say as the head lost in terms of the fluid in the pipe, and related to the velocity head.

If v = velocity in the pipe, velocity head = $v^2/2g$ and head lost = $k(v^2/2g)$ where k is a constant.

Losses of energy in a pipeline cannot be ignored. When the shock losses and friction loss have been determined they are inserted in Bernoulli's equation in the usual way.

Shock losses

10.1 Shock loss at sudden enlargement

> Show that the loss of head when a pipe undergoes a sudden increase in diameter is given by $(v_1 - v_2)^2/2g$ where v_1 is the velocity in the smaller pipe upstream of the enlargement and v_2 that in the larger pipe.
>
> A pipe increases suddenly in diameter from 0·5 m to 1 m. A mercury U-tube has one leg connected just upstream of the change and the other leg connects to the larger section a short distance downstream. If there is a difference of 35 mm in the mercury levels, the rest of the gauge being filled with water, find the discharge.

Solution. The flow will be as shown in Fig. 10.1, a region of dead water occurring as shown in which the pressure is p_0. At section (1) the pressure is p_1, velocity v_1 and area a_1. At section (2) the corresponding values are p_2, v_2 and a_2.

There is a change of momentum per second between sections (1) and (2) which is produced by the forces due to the pressures p_0, p_1, and p_2 which have a resultant opposing motion.

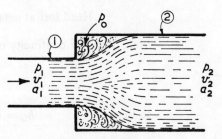

Figure 10.1

$$\text{Mass/sec flowing} = \frac{wQ}{g}$$

where $w = \rho g$ and $Q = $ discharge

$$\text{Change of velocity} = v_1 - v_2$$

therefore,

$$\text{Rate of change of momentum} = \frac{wQ}{g}(v_1 - v_2)$$

Force due to p_2 opposing motion $= p_2 a_2$

Force due to p_1 opposing motion $= p_1 a_1$

Force due to p_0 opposing motion $= -p_0(a_2 - a_1)$

Total force opposing motion $= p_2 a_2 - p_1 a_1 - p_0(a_2 - a_1)$

The value of p_0 is found experimentally to be equal to p_1, thus

$$\text{Force opposing motion} = a_2(p_2 - p_1)$$

$$= \text{rate of change of momentum}$$

$$a_2(p_2 - p_1) = \frac{wQ}{g}(v_1 - v_2)$$

or since $Q = a_2 v_2$,

$$a_2(p_2 - p_1) = \frac{wa_2 v_2}{g}(v_1 - v_2)$$

$$\frac{p_2 - p_1}{w} = \frac{v_2 v_1}{g} - \frac{v_2^2}{g} = \frac{2(v_2 v_1 - v_2^2)}{2g} \qquad . \qquad . \quad (1)$$

If $h_L = $ head lost at the enlargement, then by Bernoulli's theorem,

$$\frac{p_1}{w} + \frac{v_1^2}{2g} = \frac{p_2}{w} + \frac{v_2^2}{2g} + h_L$$

$$h_L = \frac{v_1^2 - v_2^2}{2g} - \frac{(p_2 - p_1)}{w}$$

Substituting for $(p_2 - p_1)/w$ from equation (1)

$$h_L = \frac{v_1^2 - v_2^2}{2g} - \frac{2v_2 v_1 - 2v_2^2}{2g} = \frac{v_1^2 - 2v_2 v_1 + v_2^2}{2g}$$

$$\text{Head lost at enlargement} = h_\mathrm{L} = \frac{(v_1 - v_2)^2}{2g} \qquad . \qquad . \quad (2)$$

Note. Since for continuity of flow $a_1 v_1 = a_2 v_2$,

$$v_2 = \frac{a_1}{a_2} v_1$$

so that

$$h_\mathrm{L} = \left(1 - \frac{a_1}{a_2}\right)^2 \frac{v_1^2}{2g} = k \frac{v_1^2}{2g}$$

Thus the head lost is a function of the velocity head.

If difference of mercury levels $= 35\,\text{mm}$ and there is water above the mercury,

$$\text{Difference of head} = 0{\cdot}035(13{\cdot}6 - 1) = 0{\cdot}442\,\text{m of water}$$

$$= \frac{p_2 - p_1}{w}$$

From Bernoulli's theorem, allowing for the loss at the enlargement

$$\frac{p_1}{w} + \frac{v_1^2}{2g} = \frac{p_2}{w} + \frac{v_2^2}{2g} + \frac{(v_1 - v_2)^2}{2g}$$

$$\frac{p_2 - p_1}{w} = \frac{v_1^2 - v_2^2}{2g} - \frac{(v_1 - v_2)^2}{2g}$$

$$= \frac{2v_2(v_1 - v_2)}{2g} \qquad . \qquad . \qquad . \qquad . \quad (3)$$

For continuity of flow $\quad \frac{1}{4}\pi d_1^2 v_1 = \frac{1}{4}\pi d_2^2 v_2$

Putting $d_1 = 0{\cdot}5\,\text{m}$ and $d_2 = 1{\cdot}0\,\text{m}$

$$v_2 = \left(\frac{d_1}{d_2}\right)^2$$

$$v_2 = \tfrac{1}{4} v_1$$

Substituting in equation (3)

$$\frac{p_2 - p_1}{w} = \frac{3}{8} \frac{v_1^2}{2g}$$

Putting $(p_2 - p_1)/w = 0{\cdot}442\,\text{m of water}$

$$v_1^2 = \frac{8}{3} \times 2g \times 0{\cdot}442 = 23{\cdot}1$$

$$v_1 = 4{\cdot}8\,\text{m/s}$$

$$\text{Discharge} = a_1 v_1 = \tfrac{1}{4}\pi(0{\cdot}5)^2 \times 4{\cdot}8 = \mathbf{0{\cdot}943\,m^3/s}$$

Special Case. When a pipe discharges into a large reservoir through a sharp exit, conditions are equivalent to a sudden enlargement (Fig. 10.2)

$$v_1 = \text{pipe velocity} = v$$

$$v_2 = \text{reservoir velocity} = 0$$

From the formula for a sudden enlargement, equation (3)

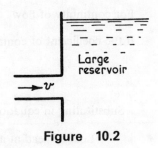

Figure 10.2

Loss of head at a sharp exit into a reservoir $= \dfrac{v^2}{2g}$

If the exit is rounded this loss is greatly reduced and is usually negligible.

10.2 Sudden contraction

> Derive an expression for the loss of head which occurs when flow passes through a sudden contraction in a pipeline. Assume that a *vena contracta* forms inside the smaller pipe and express the head lost in terms of the coefficient of contraction and the velocity in the smaller pipe.
>
> A pipe carrying $0 \cdot 06$ m³/s suddenly contracts from 200 mm to 150 mm diam. Assuming that a *vena contracta* is formed in the smaller pipe, calculate the coefficient of contraction if the pressure head at a point upstream of the contraction is $0 \cdot 655$ m greater than at a point just downstream of the *vena contracta*.

Solution. In a sudden contraction the flow converges to form a *vena contracta* at section (3) (Fig. 10.3) in the smaller pipe. The loss of energy in the convergence from sections (1) to (3) is small and the main loss occurs in the enlargement from sections (3) to (2). It is usual to ignore the loss from sections (1) to (3) and treat the loss from (3) to (2) as if it were that due to a sudden enlargement from the area of the *vena contracta* a_o to the area a_2 of the smaller pipe.

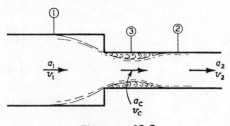

Figure 10.3

Using equation (3) of Example 10.1

$$\text{Loss of head} = \frac{(v_o - v_2)^2}{2g} \qquad . \qquad . \qquad . \quad (1)$$

For continuity of flow $\qquad a_2v_2 = a_cv_c$

If the coefficient of contraction $= C_c = a_c/a_2$

$$v_c = \frac{a_2v_2}{a_c} = \frac{1}{C_c}\,v_2$$

Substituting in equation (1)

$$\text{Loss of head at a sudden contraction} = \left(\frac{1}{C_c} - 1\right)^2 \frac{v_2{}^2}{2g}$$

Note that although the area of the larger pipe does not appear in this expression, the value of C_c depends on the ratio a_1/a_2.

Inserting this expression for the loss of head in Bernoulli's equation,

$$\frac{p_1}{w} + \frac{v_1{}^2}{2g} = \frac{p_2}{w} + \frac{v_2{}^2}{2g} + \frac{v_2{}^2}{2g}\left(\frac{1}{C_c} - 1\right)^2$$

$$\frac{p_1 - p_2}{w} = \frac{v_2{}^2}{2g}\left\{1 + \left(\frac{1}{C_c} - 1\right)^2\right\} - \frac{v_1{}^2}{2g}$$

Now
$$\frac{p_1 - p_2}{w} = 0{\cdot}655\,\text{m}$$

$$v_1 = \frac{Q}{a_1} = \frac{2}{\frac{1}{4}\pi(0{\cdot}2)^2} = 1{\cdot}91\,\text{m/s}$$

$$v_2 = \frac{Q}{a_2} = \frac{2}{\frac{1}{4}\pi(0{\cdot}15)^2} = 3{\cdot}4\,\text{m/s}$$

Thus
$$0{\cdot}655 = \frac{(3{\cdot}4)^2}{2g}\left\{1 + \left(\frac{1}{C_c} - 1\right)^2\right\} - \frac{(1{\cdot}91)^2}{2g}$$

$$12{\cdot}86 = 11{\cdot}6\left\{1 + \left(\frac{1}{C_c} - 1\right)^2\right\} - 3{\cdot}65$$

$$1 + \left(\frac{1}{C_c} - 1\right)^2 = \frac{16{\cdot}51}{11{\cdot}6} = 1{\cdot}39$$

$$\left(\frac{1}{C_c} - 1\right)^2 = 0{\cdot}39$$

$$\left(\frac{1}{C_c} - 1\right) = 0{\cdot}625, \quad \frac{1}{C_c} = 1{\cdot}625$$

Coefficient of contraction $= C_c = \mathbf{0{\cdot}615}$

Special Case. If the entrance to a pipeline from a reservoir is sharp (not rounded or bell-mouthed) it is equivalent to a sudden contraction from a pipe of infinite size to that of the pipeline. The loss of head at a sharp entrance is $\frac{1}{2}(v^2/2g)$ where v is the velocity in the pipe.

10.3 Orifice plate in pipeline

Derive an expression for the loss of head which occurs when a fluid flows through an orifice plate inserted in a pipeline if the velocity of flow in the pipe is V, the pipe area A, the orifice area a and a *vena contracta* forms downstream of the orifice with a coefficient of contraction of C_c with reference to the area of the orifice.

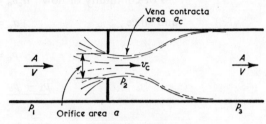

Figure 10.4

Solution. The *vena contracta* forms downstream (Fig. 10.4) with an area $a_o = C_c a$. The velocity in the *vena contracta* is v_o.

The loss of energy occurs in the enlargement from the *vena contracta* to the full bore section downstream and is treated as a sudden enlargement, so that

$$\text{Loss of head} = \frac{(v_o - V)^2}{2g}$$

For continuity of flow $a_o v_o = AV$

But, $a_o = C_c a$, so that $\quad v_o = \frac{A}{a_o} V = \frac{A}{C_c a} V$

Therefore $\qquad \text{Loss of head} = \frac{\left(\dfrac{A}{C_c a} V - V\right)^2}{2g}$

$$= \left(\frac{A}{C_c a} - 1\right)^2 \frac{V^2}{2g}$$

10.4 Orifice plate in pipeline

A diaphragm with a central hole 75 mm in diam is placed in a 150 mm diam pipe. The velocity of the water in the pipe is 0·27 m/s. Find the pressure difference between a tapping just upstream of the diaphragm and a tapping, (*a*) adjacent to the downstream face, (*b*) downstream of the *vena contracta*, given that $C_c = 0·55$.

Solution. The loss of energy occurs downstream of the *vena contracta*. In case (*a*) no loss of energy need be considered and the downstream tapping will be at the *vena contracta*.

(a) Referring to Fig. 10.4 and applying Bernoulli's equation to the upstream tapping and the *vena contracta*, assuming no loss of energy,

$$\frac{p_1}{w} + \frac{V^2}{2g} = \frac{p_2}{w} + \frac{v_c^2}{2g}$$

$$\frac{p_1 - p_2}{w} = \frac{1}{2g}(v_c^2 - V^2)$$

For continuity of flow $\quad a_c v_c = AV$

and since $\quad\quad\quad\quad\quad a_c = C_c a$

$$v_c = \frac{A}{C_c a}V$$

$$\frac{p_1 - p_2}{w} = \frac{V^2}{2g}\left\{\left(\frac{A}{C_c a}\right)^2 - 1\right\}$$

Putting $V = 0.27$ m/s, $\dfrac{A}{a} = \dfrac{\frac{1}{4}\pi \times (0.15)^2}{\frac{1}{4}\pi \times (0.075)^2} = 4$, $C_c = 0.55$

$$\frac{p_1 - p_2}{w} = \frac{(0.27)^2}{2g}\left\{\left(\frac{4}{0.55}\right)^2 - 1\right\}$$

$$\frac{p_1 - p_2}{w} = \frac{0.073 \times 52}{2g} = \mathbf{0.194\,m\ of\ water}$$

(b) Applying Bernoulli's equation to the upstream tapping and to a tapping downstream from the *vena contracta*, taking into account the shock loss.

$$\frac{p_1}{w} + \frac{V^2}{2g} = \frac{p_3}{w} + \frac{V^2}{2g} + \frac{V^2}{2g}\left\{\frac{A}{C_c a} - 1\right\}^2$$

$$\frac{p_1 - p_3}{w} = \frac{V^2}{2g}\left(\frac{1}{C_c}\cdot\frac{A}{a} - 1\right)^2$$

$$= \frac{(0.27)^2}{2g}\left(\frac{4}{0.55} - 1\right)^2$$

$$= \frac{0.073 \times (7.27 - 1)^2}{2g} = \mathbf{0.147\,m\ of\ water}$$

Friction loss

10.5 Loss due to friction, Darcy formula

Derive an expression for the loss of head due to friction in a pipeline in terms of the velocity head, assuming that the frictional resistance per unit area of pipe wall is proportional to the square of the mean velocity of flow.

Find the loss of head due to friction in a pipe 300 m long and 150 mm diam when the discharge is 2.73 m³/min and the resistance coefficient $f = 0.01$.

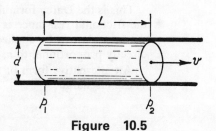

Figure   10.5

Solution. Consider a cylinder of fluid of length L (Fig. 10.5) completely filling the pipe of cross-sectional area A and moving with a mean velocity v.

The forces acting on the cylinder are the force due to pressure difference and the force due to frictional resistance. Since the velocity is constant and there is no acceleration, the resultant of these two forces in the direction of motion must be zero.

$$\text{Force due to pressure difference} = (p_1 - p_2)A$$

If $q =$ frictional resistance per unit area at unit velocity and the frictional resistance varies as v^2,

$$\text{Frictional resistance/unit area at velocity } v = qv^2$$

$$\text{Force due to friction on surface of pipe} = qv^2 \times \text{surface area}$$
$$= qv^2 \times P \times L$$

where $P =$ perimeter of the cross-section.

$$\text{Force due to pressure difference} = \text{Force due to friction}$$

$$(p_1 - p_2)A = qv^2 PL$$

If $\qquad h_f =$ head lost in friction in the length L

$$h_f = \frac{p_1 - p_2}{w} = \frac{q}{w} v^2 \frac{P}{A} L \qquad . \qquad . \qquad . \qquad . \qquad (1)$$

Or, multiplying top and bottom by $2g$,

$$h_f = \frac{2gq}{w} \frac{P}{A} L \frac{v^2}{2g}$$

Now $\qquad \dfrac{A}{P} =$ hydraulic radius of the pipe $= m$

and $2gq/w$ is a constant called the resistance coefficient and denoted by f, so that

$$h_f = \frac{fL}{m} \frac{v^2}{2g}$$

If $d =$ pipe diameter,

$$m = \frac{A}{P} = \frac{\frac{1}{4}\pi d^2}{\pi d} = \tfrac{1}{4}d$$

$$h_f = \frac{4fL}{d} \frac{v^2}{2g}$$

This is the Darcy formula for the loss of head in pipelines. Although the frictional resistance is only proportional to v^2 for turbulent flow, the formula can be applied to viscous flow, in which the resistance is directly proportional to v, by making f inversely proportional to v.

An alternative form of the Darcy formula is sometimes useful. If Q is the discharge

$$v = \frac{Q}{A} = \frac{4Q}{\pi d^2}$$

$$h_f = \frac{4fL}{d} \cdot \frac{v^2}{2g} = \frac{64fLQ^2}{2g\pi^2 d^5} = \frac{fLQ^2}{3 \cdot 03 d^5} \text{ working in SI units}$$

$$h_f = \frac{fLQ^2}{3d^5}$$

within one per cent error.

If $Q = 2 \cdot 73 \, \text{m}^3/\text{min} = 0 \cdot 0455 \, \text{m}^3/\text{s}, f = 0 \cdot 01, L = 300 \, \text{m}, d = 150 \, \text{mm} = 0 \cdot 15 \, \text{m}$

$$h_f = \frac{0 \cdot 01 \times 300 \times (0 \cdot 0455)^2}{3 \times (0 \cdot 15)^5} = \textbf{27} \cdot \textbf{3 m of water}$$

10.6 Chezy formula

Derive the Chezy formula for the mean velocity v in a pipeline in terms of the hydraulic radius m and the loss of head per unit length i.

Find the loss of head due to friction in a 75 mm pipe 30 m long, for which the Chezy coefficient $C = 54 \cdot 6$ SI units, if the mean velocity of flow is $1 \cdot 8$ m/s.

Solution. The Chezy and Darcy formulae are derived in the same way and are alternatives. Equation (1) of Example 10.5 is

$$h_f = \frac{q}{w} v^2 \frac{P}{A} L$$

$$v^2 = \frac{w}{q} \frac{A}{P} \frac{h_f}{L}$$

Loss of head per unit length $= i = \dfrac{h_f}{L}$

and

$$A = \frac{m}{P}$$

$\therefore$

$$v^2 = \frac{w}{q} mi$$

and

$$v = \sqrt{\frac{w}{q}} \sqrt{(mi)}$$

This is written $v = C\sqrt{(mi)}$ where C is the Chezy coefficient.

The relation between the Darcy coefficient f which is a pure number and Chezy coefficient is that $C = \sqrt{(2g/f)}$ and it has units of $\text{m}^{1/2}/\text{s}$ in SI units.

Since $v = C\sqrt{(mi)}$

$$i = \frac{v^2}{C^2 m}$$

Loss of head $h_f = iL = \dfrac{v^2 L}{C^2 m}$

Putting $v = 1{\cdot}8$ m/s, $L = 30$ m, C $= 54.6$ SI units,

$$m = \tfrac{1}{4}d = \tfrac{1}{4} \times 0{\cdot}075 \text{m}$$

Loss of head $h_f = \dfrac{(1{\cdot}8)^2 \times 30}{(54{\cdot}6)^2 \times \tfrac{1}{4} \times 0{\cdot}075} = \mathbf{1{\cdot}75m}$

10.7 Manning formula and Hazen-Williams formula

What are the defects of the Darcy and Chezy forumlae for the estimation of the loss of head in a pipeline under turbulent flow conditions? State two empirical formulae which are intended to allow for these defects.

Solution. The Darcy and Chezy formulae are based on the same reasoning and therefore have the same defects. It was assumed in Example 10.5 that the frictional resistance varied as the square of the mean velocity for turbulent flow and thus

$$h_f = \frac{4fL}{d}\frac{v^2}{2g}$$

Since both h_f and $v^2/2g$ are heads and L/d is a ratio, f will be a pure number.

In practice the loss of head with turbulent flow does not vary as the square of the velocity, but as some power varying from 1·7 to 2 or more. Thus f must vary as some power of the velocity and will not be a constant for a given pipe for all rates of flow.

Also, the value of f is dependent on the roughness of the pipe surface, and may vary for steep slopes compared with flat slopes.

To overcome these defects the following formulae have been devised which are modifications of the Chezy formula $v = C\sqrt{(mi)}$.

Manning formula: $v = \dfrac{1}{n} m^{2/3} i^{1/2}$ in SI units

where n is a roughness coefficient increasing with the roughness of the pipe from 0·009 for glass to 0·022 for dirty cast-iron pipe.

Hazen-Williams formula: $v = 0{\cdot}82 C_1 m^{0{\cdot}6} i^{0{\cdot}54}$ in SI units

where C_1 is a coefficient varying from 140 for very smooth pipes to 80 or less for old iron pipes in bad condition.

Problems

1 A $0 \cdot 3$ m diam pipe through which water flows at the rate of $0 \cdot 282$ m³/s suddenly enlarges to $0 \cdot 6$ m in diam. If the axis of the pipe is horizontal and the water in a vertical tube connected to the larger pipe stands $0 \cdot 36$ m higher than the level in a tube connected to the smaller pipe, determine the coefficient K if the shock loss is expressed as $Kv^2/2g$, where v is the velocity in the smaller pipe.
Answer $0 \cdot 496$

2 A 100 mm diam pipe carrying $1 \cdot 8$ m³/min of water suddenly enlarges to 150 mm diam. Find (*a*) the loss of head due to the sudden enlargement, (*b*) the difference in pressure in kN/m² in the two pipes taken between points just outside the area of disturbance due to change of section, (*c*) the corresponding difference in pressure if the change in diameter were to be effected by a gradual cone (assume no loss). The axis of the pipe is horizontal.
Answer $0 \cdot 229$ m; $3 \cdot 6$ kN/m²; $5 \cdot 85$ kN/m²

3 Water flows down a straight inclined pipe. At a point A, 45 m above datum, the section of the pipe suddenly doubles in area. The pressure in the smaller pipe at A is 860 kN/m² and the velocity of flow in the larger pipe is $2 \cdot 4$ m/s. Find the pressure in kN/m² at a point B which is $2 \cdot 5$ m above datum.
Answer 1283 kN/m²

4 A pipe, $0 \cdot 093$ m² in area, carries a discharge of $0 \cdot 283$ m³/s of water. If it enlarges suddenly to $0 \cdot 372$ m² and if the pressure in the smaller section is $4 \cdot 8$ kN/m², find (*a*) the head lost, (*b*) the pressure in the larger part, (*c*) the power needed to force the water through the enlargement.
Answer $0 \cdot 265$ m; $6 \cdot 53$ kN/m²; 737 W

5 Derive an expression for the loss of head at a sudden enlargement in a pipe. Water flows in a 150 mm diam pipe, and at a sudden enlargement the loss of head is found to be one-half of the velocity head in the 150 mm pipe. Determine the diameter of the enlarged portion.
Answer 277 mm

6 Derive an expression for the loss of head at a sudden contraction. A pipe carrying $0 \cdot 056$ m³/s suddenly changes diameter from (*a*) 200 mm to 150 mm, (*b*) 300 mm to 150 mm, (*c*) 450 mm to 150 mm. Find the loss of head and the pressure difference across the contraction in each case. $C_c = 0 \cdot 62$.
Answer $0 \cdot 19$ m, $0 \cdot 54$ m; $0 \cdot 19$ m, $0 \cdot 673$ m; $0 \cdot 19$ m, $0 \cdot 699$ m

7 A pipeline carrying $0 \cdot 236$ m³/s is reduced suddenly from 450 mm to 300 mm diam. Calculate the change in (*a*) total energy head, (*b*) pressure energy head. Take $C_c = 0 \cdot 67$.
Answer $0 \cdot 135$ m fall; $0 \cdot 589$ m fall

8 A 150 mm diam pipe suddenly contracts to 100 mm diam. If the flow is $1 \cdot 8$ m³/min, find (*a*) the loss of energy due to the sud-

den contraction, (*b*) the change in pressure in the two pipes, (*c*) the change in pressure head if there were no loss of energy, the pipes being joined by a gradual cone instead of a sudden contraction. Take a coefficient of contraction of 0·66.
Answer   0·1974 Nm/N; 7·79 kN/m²; 5·86 kN/m²

9   Water flows at the rate of 0·028 m³ through a 150 mm diam pipe which reduces suddenly to 100 mm diam. Find the loss of head and the difference of pressure head between tapping points either side of the contraction if the coefficient of contraction is 0·62. Assume the full loss due to the enlargement of the streamlines to be included, but make no allowance for pipe friction.

If a plate, with an orifice of diam 75 mm, is inserted at the contraction, what would then be the values of the above quantities, again taking $C_c = 0·62$?
Answer   0·258 m; 0·812 m; 2·19 m; 2·74 m

10   Water flows vertically downward through a 150 mm diam pipe with a velocity of 2·4 m/s. The pipe suddenly enlarges to 300 mm in diam. Find the loss of head and also the loss of head if the flow is reversed, the coefficient of contraction now being 0·62.
Answer   0·165 m; 0·110 m

11   Deduce an expression for the loss of head due to a sudden enlargement in a pipe conveying water. State the assumptions made.

A pipeline conveying a given quantity of water per second is suddenly changed in section. Find the ratio of the pipe diameters if the loss of head at the change in section is independent of the direction of flow. Assume a value of 0·61 for the coefficient of contraction.
Answer   1 to 1·667

12   A diaphragm with a 150 mm diam opening is inserted in a 300 mm diam pipe. If the velocity of flow is 0·6 m/s, calculate the loss of head due to the diaphragm assuming that $C_c = 0·64$.
Answer   0·507 m

13   The passage of water through a 150 mm diam pipe is restricted by a diaphragm with a 50 mm diam hole in its centre. The loss of head at the diaphragm is 0·375 m when the velocity in the pipe is 0·177 m/s. Find the value of the coefficient of contraction of the stream passing through the diaphragm.
Answer   0·552

14   Determine the loss of head due to friction in a new cast-iron pipe 360 m long and 150 mm diam which carries 42 dm³/s. Use the Darcy formula, taking $f = 0·005$.
Answer   13·82 m

15   Using the Chezy formula, find the loss of head in a circular pipe 120 m long and 75 mm diam when the velocity of flow is 4·8 m/s. Take C = 54·6 SI units.
Answer   49·5 m

16 Water is discharged from a reservoir through a pipe 1200 m long which is 400 mm diam for 600 m and 250 mm diam for the rest. Calculate the flow, taking only friction into account, if the far end of the pipe is 30 m below the reservoir level: f for the 400 mm pipe is $0 \cdot 004$ and for the 240 mm pipe $0 \cdot 006$.
Answer $0 \cdot 151$ m³/s

17 Two reservoirs whose difference of level is $13 \cdot 5$ m are connected by a pipe ABC whose highest point B is $1 \cdot 5$ m below the level in the upper reservoir A. The portion AB has a diameter of 200 mm and the portion BC a diameter of 150 mm, the friction coefficient for each being $0 \cdot 005$. The total length of the pipe is 3000 m. Find the maximum allowable length of the portion AB if the pressure head at B is not to be more than 3 m below atmospheric pressure. Neglect velocity head in the pipe, loss of head at pipe entry and loss of head at change of diameter.
Answer   2038 m

18 A head loss of $0 \cdot 755$ m is measured across an abrupt enlargement from a 75 mm diam section to a larger diameter when the flow rate is $28 \cdot 3$ dm³/s.

Find the head loss if the flow has the same magnitude but is reversed in direction. Assume that the coefficient of contraction C_c varies with the ratio of pipe areas as follows:

A_2/A_1	$0 \cdot 1$	$0 \cdot 3$	$0 \cdot 5$	$0 \cdot 7$
C_c	$0 \cdot 61$	$0 \cdot 63$	$0 \cdot 67$	$0 \cdot 73$

Answer $0 \cdot 62$ m

11

Pipeline problems

All pipeline problems should be solved by applying Bernoulli's theorem between points for which the total energy is known and including expressions for any loss of energy due to shock or to friction, thus

$$\frac{p_1}{w} + \frac{v_1^2}{2g} + z_1 = \frac{p_2}{w} + \frac{v_2^2}{2g} + z_2 + \text{shock loss} + \text{frictional loss}.$$

The standard shock losses at entry and at exit occur only when the entry or exit is "sharp" and flow is out of or into a reservoir so that conditions are those of a sudden contraction or enlargement.

The Darcy formula $\qquad h_f = \dfrac{4fL}{d} \dfrac{v^2}{2g}$

is most convenient for the frictional loss. All the terms in Bernoulli's equation represent energy per unit weight, as do the expressions for shock loss and friction loss.

When pipes are in series the frictional losses are additive. For pipes in parallel or for branching pipelines the energy equation can be written down for each branch or route without reference to alternative routes since the terms are energy per unit weight and are therefore unaffected by divisions of the flow.

Single pipe problems

11.1 Single pipe connecting reservoirs

A pipeline connecting two reservoirs having a difference of level of 6 m is 720 m long, and rises to a height of 3 m above the upper reservoir at a distance of 240 m from the entrance before falling to the lower reservoir. If the pipe is 1·2 m in diam and the frictional coefficient $f = 0·01$, what will be the discharge and the pressure at the highest point of the pipeline?

Solution. Fig. 11.1 shows the layout. Since no information is given that exit or entry to the pipe is sharp no shock loss need be considered.

Apply Bernoulli's equation to A and B on the free surface of the reservoirs where the velocities are zero and the pressures atmospheric. Take B as datum level.

Total energy per unit wt at A = Total energy per unit wt at B
+ frictional loss

$$H + \frac{p_A}{w} + \frac{v_A{}^2}{2g} = 0 + \frac{p_B}{w} + \frac{v_B{}^2}{2g} + \frac{4fL}{d}\frac{V^2}{2g}$$

$p_A = p_B$ and $v_A = v_B = 0$, so that

$$H = \frac{4fL}{d}\frac{V^2}{2g}$$

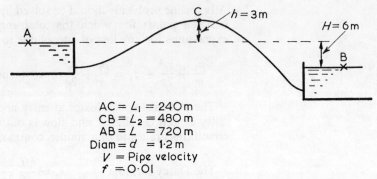

AC = L_1 = 240 m
CB = L_2 = 480 m
AB = L = 720 m
Diam = d = 1·2 m
V = Pipe velocity
f = 0·01

Figure 11.1

Putting $H = 6\,\mathrm{m}$, $f = 0{\cdot}01$, $L = 720\,\mathrm{m}$, $d = 1{\cdot}2\,\mathrm{m}$

$$6 = \frac{4 \times 0{\cdot}01 \times 720}{1{\cdot}2}\frac{V^2}{2g}$$

$$V^2 = \frac{6 \times 1{\cdot}2 \times 2 \times 9{\cdot}81}{4 \times 0{\cdot}01 \times 720} = 4{\cdot}92$$

$$V = 2{\cdot}22\,\mathrm{m/s}$$

Discharge = area × velocity = $\frac{1}{4}\pi(1{\cdot}2)^2 \times 2{\cdot}22 = \mathbf{2{\cdot}51\,m^3/s}$

To find the pressure p_C at C apply Bernoulli's theorem to A and C, taking datum level at A and $v_A = 0$

$$\frac{p_A}{w} = \frac{p_C}{w} + h + \frac{V^2}{2g} + \frac{4fL_1}{d}\frac{V^2}{2g}$$

$$\frac{p_C}{w} = \frac{p_A}{w} - h - \frac{V^2}{2g}\left(1 + \frac{4fL_1}{d}\right)$$

Putting p_A = atmospheric pressure = 0, $h = 3\,\mathrm{m}$, $f = 0{\cdot}01$, $L_1 = 240\,\mathrm{m}$, $V = 2{\cdot}22\,\mathrm{m}$, $d = 1{\cdot}2\,\mathrm{m}$

$$\frac{p_C}{w} = 0 - 3 - \frac{(2{\cdot}22)^2}{2g}\left(1 + \frac{4 \times 0{\cdot}01 \times 240}{1{\cdot}2}\right)$$

$$= -3 - \frac{4{\cdot}92 \times 9}{2g} = -5{\cdot}26\,\mathrm{m\ of\ water}$$

$$p_C = -9{\cdot}81 \times 10^3 \times 5{\cdot}26 = \mathbf{-51{\cdot}6\,kN/m^2}$$

11.2 Discharge to atmosphere

Water from a large reservoir is discharged to atmosphere through a 100 m diam pipe 450 m long. The entry from the reservoir to the pipe is sharp and the outlet is 12 m below the surface level in the reservoir. Taking $f = 0.01$ in the Darcy formula, calculate the discharge.

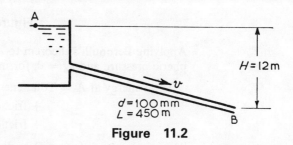

Figure 11.2

Solution. Apply Bernoulli's theorem to A and B (Fig. 11.2), assuming velocity at A is zero and that $p_A = p_B =$ atmospheric pressure.

Total energy at A = Total energy at B + loss at entry + frictional loss.

$$H = \frac{v^2}{2g} + \frac{1}{2}\frac{v^2}{2g} + \frac{4fL}{d} \cdot \frac{v^2}{2g}$$

Putting $H = 12$ m, $f = 0.01$, $L = 450$ m, $d = 100$ mm $= 0.1$ m

$$12 = \frac{v^2}{2g}\left(1.5 + \frac{4 \times 0.01 \times 450}{0.1}\right) = 181.5\frac{v^2}{2g}$$

$$v^2 = \frac{12 \times 2g}{181.5} = 1.3, \quad v = 1.14 \text{m/s}$$

Discharge $= \frac{1}{4}\pi d^2 v = \frac{1}{4}\pi(0.1)^2 \times 1.14 = \mathbf{8.96 \times 10^{-3} m/s}$

11.3 Pipes in series

Water is discharged from a reservoir into the atmosphere through a pipe 39 m long. There is a sharp entrance to the pipe and the diameter is 50 mm for 15 m from the entrance. The pipe then enlarges suddenly to 75 mm in diam for the remainder of its length. Taking into account the loss of head at entry and at the enlargement, calculate the difference of level between the surface of the reservoir and the pipe exit which will maintain a flow of 2·8 dm³/s. Take f as 0·0048 for the 50 mm pipe and 0·0058 for the 75 mm pipe.

Solution. Fig. 11.3 shows the arrangement. If $Q =$ discharge

$$Q = \frac{1}{4}\pi d_1^2 v_1 = \frac{1}{4}\pi d_2^2 v_2$$

$$v_1 = \frac{4Q}{\pi d_1^2} = \frac{4 \times 2.8 \times 10^{-3}}{\pi \times (0.05)^2} = 1.426 \text{m/s}$$

$$v_2 = \frac{4Q}{\pi d_2^2} = \frac{4 \times 2.8 \times 10^{-3}}{\pi \times (0.075)^2} = 0.634 \text{m/s}$$

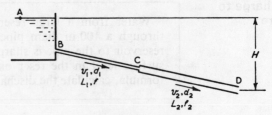

Figure 11.3

Applying Bernoulli's theorem to A and D at which $p_A = p_D$ = atmospheric pressure and $v_A = 0$, for unit weight

Total energy at A = Total energy at D + shock loss at B

+ frictional loss in BC + shock loss at C

+ frictional loss in CD

There is no shock loss at D as discharge is to atmosphere.

$$H = \frac{v_2^2}{2g} + \frac{1}{2} \cdot \frac{v_1^2}{2g} + \frac{4f_1L_1}{d_1} \frac{v_1^2}{2g} + \frac{(v_1 - v_2)^2}{2g} + \frac{4f_2L_2}{d_2} \frac{v_2^2}{2g}$$

$$= \frac{v_1^2}{2g} \left(\frac{1}{2} + \frac{4f_1L_1}{d_1} \right) + \frac{v_2^2}{2g} \left(1 + \frac{4f_2L_2}{d_2} \right) + \frac{(v_1 - v_2)^2}{2g}$$

$$= \frac{(1 \cdot 426)^2}{2g} \left(\frac{1}{2} + \frac{4 \times 0 \cdot 0048 \times 15}{0 \cdot 05} \right)$$

$$+ \frac{(0 \cdot 634)^2}{2g} \left(1 + \frac{4 \times 0 \cdot 0058 \times 24}{0 \cdot 075} \right) + \frac{(0 \cdot 792)^2}{2g}$$

$$= 0 \cdot 647 + 0 \cdot 173 + 0 \cdot 032 \, \text{m}$$

Difference of level = H = **0·852 m of water**

11.4 Uniform draw off

> A water main is 100 mm in diam and 4·8 km long. Supplies are taken from it uniformly at 7·5 litres/h per metre of its length. Calculate the greatest difference of head between the point of admission to the main and a point of supply when admission is (*a*) at one end, and (*b*) at the centre of the main: $f = 0 \cdot 006$.

Solution. With continuous draw-off along the pipeline the velocity and therefore the frictional loss of head per unit length will vary from point to point. The problem is solved by considering a short length of pipe and then integrating.

At a distance x from the inlet let the velocity in the pipe be v and the head lost in a distance δx from the point be δh. Then by the Darcy formula

$$\delta h = \frac{4f\delta x}{D} \cdot \frac{v^2}{2g}$$

where D = pipe diameter.

Let Q_1 = total rate of input, K = draw-off per unit length and Q = rate of flow past the given point. Since

$$\text{Total input} = \text{Total draw off}$$

$Q = 0$ when $x = L$ and $Q_1 = KL$. Thus

$$Q = K(L - x)$$

$$v = \frac{4Q}{\pi D^2} = \frac{4K}{\pi D^2}(L - x)$$

$$\delta h = \frac{4f}{2gD} \cdot \frac{16K^2}{\pi^2 D^4}(L - x)^2 \delta x$$

Loss of head between input and extreme end

$$= h_1 - h_2 = h = \frac{64fK^2}{2g\pi^2 D^5}\int_{x=L}^{x=0}(L - x)^2 dx$$

$$h = \frac{64fK^2L^3}{6g\pi^2 D^5} = \frac{0{\cdot}11fK^2L^3}{D^5}$$

Putting $f = 0{\cdot}006$,

$$K = 7{\cdot}5\,\text{litres/hr/m} = \frac{7{\cdot}5 \times 10^{-3}}{3600} = 2{\cdot}085 \times 10^{-6}\,\text{m}^3/\text{s/m}$$

$$D = 0{\cdot}1\,\text{m}$$

$$h = \left[\frac{0{\cdot}11 \times 0{\cdot}006 \times (2{\cdot}085 \times 10^{-6})^2 L^3}{(0{\cdot}1)^5}\right] = 2{\cdot}87 \times 10^{-10}L^3\,\text{m}$$

(a) When input is at one end $L = 4{\cdot}8 \times 10^3\,\text{m}$

Difference of head $= h = 2{\cdot}87 \times 10^{-10} \times (4{\cdot}8)^3 \times 10^9$

$$= 31\,\text{m}$$

(b) For input at the centre $L = 2{\cdot}4 \times 10^3\,\text{m}$

Difference of head $= h = 2{\cdot}87 \times 10^{-10} \times (2{\cdot}4)^3 \times 10^9$

$$= 3{\cdot}88\,\text{m}$$

11.5 Improving flow

The outlet from a water tank consists of a sharp-edged entry to a horizontal pipe of 150 mm diam, 30 m long, which discharges to atmosphere through a valve. When fully open the valve causes a head loss of $\frac{1}{4}(v2/2g)$. The pipe has a Darcy friction coefficient of $f = 0{\cdot}006$. What are the limiting improvements (per cent) obtainable (a) by the replacement of the abrupt entry by a carefully rounded entry, and (b) by the removal of the valve, in terms of the increase on the discharge.

Solution. Referring to Fig. 11.4, let the loss of energy/unit wt at the entry be h_{L1} and at the valve be h_{L2}.

Applying Bernoulli's equation to the free surface and the outlet, for

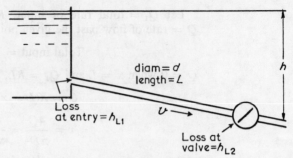

Figure 11.4

unit weight,

Total energy at free surface

= Total energy at outlet + entry loss + friction loss + valve loss

$$h = \frac{v^2}{2g} + h_{L1} + h_f + h_{L2} \qquad . \qquad . \qquad . \qquad (1)$$

For the original case before improvement $v = v_1$.

For an abrupt entry $\qquad\qquad h_{L1} = 0.5 \frac{v_1^2}{2g}$

$$h_{L2} = 0.25 \frac{v_1^2}{2g}$$

$$h_f = \frac{4fL}{d} \frac{v_1^2}{2g} = \frac{4 \times 0.006 \times 30}{0.15} = 4.8 \frac{v_1^2}{2g}$$

Substituting in (1)

$$h = \frac{v_1^2}{2g} + 0.5 \frac{v_1^2}{2g} + 4.8 \frac{v_1^2}{2g} + 0.25 \frac{v_1^2}{2g} = 6.55 \frac{v_1^2}{2g}$$

$$v_1 = \sqrt{\frac{2gh}{6.55}} = 0.3906\sqrt{(2gh)}$$

$$Q_1 = av_1 = 0.3906a\sqrt{(2gh)}$$

(a) If the entry is rounded, the entry loss is negligible so that $h_{L1} = 0$. Substituting in equation (1), putting $v = v_2$,

$$h = \frac{v_2^2}{2g} + 0 + 4.8 \frac{v_2^2}{2g} + 0.25 \frac{v_2^2}{2g} = 6.05 \frac{v_2^2}{2g}$$

where v_2 is the new velocity in the pipe.

$$v_2 = \sqrt{\frac{2gh}{6.05}} = 0.4065\sqrt{(2gh)}$$

New discharge $Q_2 = av_2 = 0.4065a\sqrt{(2gh)}$

Percentage increase $= \dfrac{Q_2 - Q_1}{Q_1} \times 100$

$$= \frac{(0.4065 - 0.3906)a\sqrt{(2gh)}}{0.3906a\sqrt{(2gh)}} \times 100$$

$$= \frac{1 \cdot 59}{0 \cdot 3906} = 4 \cdot 07 \text{ per cent}$$

(b) With the abrupt entry but with the valve removed,

$$h_{L1} = 0 \cdot 5 \frac{v_3^2}{2g}, \quad h_{L2} = 0$$

Substituting in equation (1), putting $v = v_3$,

$$h = \frac{v_3^2}{2g} + 0 \cdot 5 \frac{v_3^2}{2g} + 4 \cdot 8 \frac{v_3^2}{2g} + 0 = 6 \cdot 3 \frac{v_3^2}{2g}$$

where v_3 is the new velocity in the pipe.

$$v_3 = \sqrt{\frac{2gh}{6 \cdot 3}} = 0 \cdot 3984 \sqrt{(2gh)}$$

$$Q_3 = av_3 = 0 \cdot 3984 a\sqrt{(2gh)}$$

$$\text{Percentage increase} = \frac{Q_3 - Q_1}{Q_1} \times 100$$

$$= \frac{(0 \cdot 3984 - 0 \cdot 3906)}{0 \cdot 3906} a\sqrt{(2gh)} \times 100$$

$$= \frac{0 \cdot 78}{0 \cdot 3906} = 2 \text{ per cent}$$

11.6 Hydraulic gradient

Two reservoirs are connected by a pipeline which is 150 mm in diam for the first 6 m and 225 mm in diam for the remaining 15 m. The entrance and exit are sharp and the change of section is sudden. The water surface in the upper reservoir is 6 m above that in the lower. Tabulate the losses of head which occur and calculate the rate of flow in m³/s. Friction coefficient f is 0·01 for both pipes.

Draw also the hydraulic gradient and the total energy gradient.

Solution. The arrangement is shown in Fig. 11.5. The velocities v_1 and v_2 are related by the continuity of flow equation

$$\tfrac{1}{4}\pi d_1^2 v_1 = \tfrac{1}{4}\pi d_2^2 v_2$$

Since $d_1 = 150 \text{mm}$ and $d_2 = 225 \text{mm}$

$$v_1 = \left(\frac{225}{150}\right)^2 v_2 = \left(\frac{9}{4}\right) v_2$$

The losses are

$$\text{Loss at entry} = \frac{1}{2} \frac{v_1^2}{2g} = \frac{1}{2} \left(\frac{9}{4}\right)^2 \frac{v_2^2}{2g} = 2 \cdot 53 \frac{v_2^2}{2g}$$

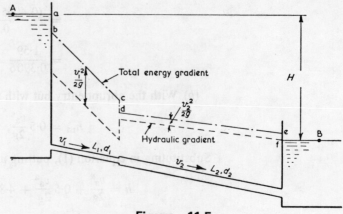

Figure 11.5

Friction in the 6-in. pipe

$$= \frac{4fL_1}{d_1} \frac{v_1{}^2}{2g} = \frac{4 \times 0 \cdot 01 \times 6}{0 \cdot 15} \frac{v_1{}^2}{2g}$$

$$= 1 \cdot 6 \frac{v_1{}^2}{2g} = 1 \cdot 6 \left(\frac{9}{4}\right)^2 \frac{v_2{}^2}{2g} = 8 \cdot 10 \frac{v_2{}^2}{2g}$$

$$\text{Shock at enlargement} = \frac{(v_1 - v_2)^2}{2g} = \frac{v_2{}^2}{2g} \left(\frac{9}{4} - 1\right)^2 = 1 \cdot 56 \frac{v_2{}^2}{2g}$$

$$\text{Friction in the 9-in. pipe} = \frac{4fL_2}{d_2} \frac{v_2{}^2}{2g} = \frac{4 \times 0 \cdot 01 \times 15}{0 \cdot 225} \frac{v_2{}^2}{2g} = 2 \cdot 67 \frac{v_2{}^2}{2g}$$

$$\text{Shock at exit} = \frac{v_2{}^2}{2g} = 1 \cdot 00 \frac{v_2{}^2}{2g}$$

$$\text{Total loss of head} = 15 \cdot 86 \frac{v_2{}^2}{2g}$$

Applying Bernoulli's equation to A and B for unit weight

Total energy at A = Total energy at B + Losses

Pressures at A and B are equal and if the reservoirs are large the velocities will be zero. Taking datum level at B

$$H = 0 + \text{Losses}$$

$$6 = 15 \cdot 86 \frac{v_2{}^2}{2g}$$

$$\therefore \qquad v_2 = \sqrt{\frac{6 \times 2g}{15 \cdot 86}} = 2 \cdot 72 \,\text{m/s}$$

$$\text{Discharge} = \tfrac{1}{4}\pi d_2{}^2 v_2 = \tfrac{1}{4}\pi (0 \cdot 225)^2 \times 2 \cdot 72 = \mathbf{0 \cdot 185 m^3/s}$$

The total energy at any point can be represented graphically, as shown in Fig. 11.5. At the entry to the pipeline there will be an entry loss *ab*; next follows the frictional loss in the first pipe, which gives the sloping straight line *bc*. The loss at the sudden enlargement is shown as *cd*. From *d* to *e* the frictional loss in the larger pipe causes the total energy to fall slowly and the shock loss at exit *ef* brings the total energy gradient to the level of the surface in the lower reservoir.

The hydraulic gradient is obtained by plotting the sum of the potential and pressure energy and will therefore be a distance equal to the velocity head in the pipe below the total energy gradient. The hydraulic gradient shows the level to which the liquid in the pipe would rise if a vertical standpipe was inserted in the pipeline at the point under consideration.

Parallel pipes and branching pipelines

11.7 Parallel pipes

> Two reservoirs are connected by three pipes laid in parallel, their diameters are respectively d, $2d$ and $3d$ and they are all the same length L. Assuming f to be the same for all the pipes, what will be the discharge through the larger pipes if that through the smallest is $0\cdot03$ m³/s.

Solution. Since the pressures at entry and exit of each pipe are unaffected by the flow in the other pipes, each pipe can be treated separately, using the whole available difference of head in each case.

If Q_1, Q_2 and Q_3 are the discharges in the pipes of diameter d, $2d$ and $3d$ respectively, and H is the difference of level of the reservoirs,

$$H = \frac{fLQ_1^2}{3d_1^5} = \frac{fLQ_2^2}{3d_2^5} = \frac{fLQ_3^2}{3d_3^5}$$

Putting $Q_1 = 0\cdot03$ m³/s, $d_1 = d$, $d_2 = 2d$, $d_3 = 3d$

$$\frac{(0\cdot03)^2}{d^5} = \frac{Q_2^2}{(2d)^5} = \frac{Q_3^2}{(3d)^5}$$

$$Q_2 = 0\cdot03\sqrt{32} = \mathbf{0\cdot1695\,m^3/s}$$

$$Q_3 = 0\cdot03\sqrt{243} = \mathbf{0\cdot459\,m^3/s}$$

11.8 Single and parallel pipes in series

> Two reservoirs have a difference of level of 6 m and are connected by a pipeline which consists of a single 600 mm diam pipe 3000 m long feeding a junction from which two pipes, each of 300 mm diam and 3000 m long, lead in parallel to the lower reservoir. If $f = 0\cdot01$, what will be the total discharge?

Solution. Considering a unit weight of the fluid, it may flow from the upper to the lower reservoir either through pipes BC and CD (Fig. 11.6) or through pipes BC and CE. By either route the head producing flow is *H* which is the energy per unit weight available.

Applying Bernoulli's theorem to points A and F on the free surfaces:

For unit weight flowing through BCD

$$z_A = z_F + \frac{4fL_1}{d_1}\frac{v_1^2}{2g} + \frac{4fL_2}{d_2}\frac{v_2^2}{2g}$$

$$z_A - z_F = H = \frac{4fL_1}{d_1}\frac{v_1^2}{2g} + \frac{4fL_2}{d_2}\frac{v_2^2}{2g} \qquad . \quad . \quad (1)$$

For unit weight flowing through BCE

$$z_A - z_F = H = \frac{4fL_1}{d_1}\frac{v_1^2}{2g} + \frac{4fL_3}{d_3}\frac{v_3^2}{2g} \qquad . \quad . \quad (2)$$

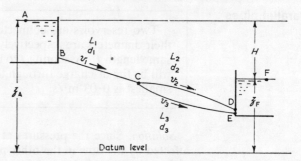

Figure 11.6

From equations (1) and (2),

$$\frac{4fL_2}{d_2}\frac{v_2^2}{2g} = \frac{4fL_3}{d_3}\frac{v_3^2}{2g}$$

But $L_2 = L_3$ and $d_2 = d_3$, so that $v_2 = v_3$

Also for continuity of flow

$$\tfrac{1}{4}\pi d_1^2 v_1 = \tfrac{1}{4}\pi d_2^2 v_2 + \tfrac{1}{4}\pi d_3^2 v_3$$

Putting $v_3 = v_2$ and $d_3 = d_2$

$$v_2 = \tfrac{1}{2}\left(\frac{d_1}{d_2}\right)^2 v_1 = \tfrac{1}{2}\left(\frac{2}{1}\right)^2 v_1 = 2v_1$$

Substituting in equation (1)

$$H = \frac{4fL_1}{d_1}\frac{v_1^2}{2g} + \frac{4fL_2}{d_2}\frac{4v_1^2}{2g}$$

$$= \frac{4fv_1^2}{2g}\left(\frac{L_1}{d_1} + \frac{4L_2}{d_2}\right)$$

$$6 = \frac{4 \times 0.01}{2g} v_1^2 \left(\frac{3000}{0.6} + \frac{4 \times 3000}{0.3} \right)$$

$$v_1^2 = \frac{6 \times 2g}{4 \times 0.01 \times 45000} = 0.0656$$

$$v_1 = 0.256 \,\text{m/s}$$

$$\text{Discharge} = \tfrac{1}{4}\pi d_1^2 v_1 = \tfrac{1}{4}\pi(0.6)^2 \times 0.256 = \mathbf{0.0725 \,m^3/s}$$

11.9 Tapped pipeline

Two reservoirs having a constant difference in water level of 66 m are connected by a 225 mm diam pipe, 4 km long. The pipe is tapped at a point distant 1·6 km from the upper reservoir, and water is drawn off at the rate of 42·5 dm³/s.

If the friction coefficient f is 0·009, determine the rate in dm³/s at which water enters the lower reservoir, neglecting all losses except pipe friction. Sketch the hydraulic gradient for the pipe.

Solution. If C is the point at which the pipe is tapped (Fig. 11.7), Q_1 the rate of flow in BC, Q_2 in CD and Q_3 the rate of draw-off:

For continuity of flow

$$Q_1 = Q_2 + Q_3 \qquad \qquad . \qquad . \qquad . \qquad . \quad (1)$$

Applying Bernoulli's theorem to A and E for unit weight

Total energy at A = Total energy at E + loss in BC + loss in CD

Taking datum level at E

$$H = \frac{fL_1 Q_1^2}{3d^5} + \frac{fL_2 Q_2^2}{3d^5} = \frac{f}{3d^5}(L_1 Q_1^2 + L_2 Q_2^2)$$

$$66 = \frac{0.009}{3 \times (0.225)^5}(1.6 \times 10^3 Q_1^2 + 2.4 \times 10^3 Q_2^2)$$

$$Q_1^2 + 1.5 Q_2^2 = \frac{66 \times 3 \times (0.225)^5}{0.009 \times 1.6 \times 10^3} = 7.96 \times 10^{-3} \quad . \quad (2)$$

From (1) $\qquad Q_1 = Q_2 + 42.5 \times 10^{-3}$

and $\qquad Q_1^2 = Q_2^2 + 85 \times 10^{-3} Q_2 + 1.81 \times 10^{-3}$

Substituting in (2)

$$2.5 Q_2^2 + 85 \times 10^{-3} Q_2 + 1.81 \times 10^{-3} = 7.96 \times 10^{-3}$$

$$2.5 Q_2^2 + 85 \times 10^{-3} - 6.15 \times 10^{-3} = 0$$

$$Q_2 = \frac{-85 \times 10^{-3} \pm \sqrt{(7225 \times 10^{-6} + 61.5 \times 10^{-3})}}{5}$$

$$= \mathbf{35.4 \times 10^{-3} \, dm^3/s}$$

$$Q_1 = Q_2 + 42{\cdot}5 \times 10^{-3} = 77{\cdot}9 \times 10^{-3} \, \text{dm}^3/\text{s}$$

$$\text{Loss of head in BC} = \frac{fLQ_1^2}{3d^5}$$

$$= \frac{0{\cdot}009 \times 1{\cdot}6 \times 10^3 \times (77{\cdot}9 \times 10^{-3})^2}{3 \times (0{\cdot}225)^5} = 50{\cdot}8$$

$$\text{Loss of head in CD} = 66 - 50{\cdot}8 = \mathbf{15{\cdot}2m}$$

The hydraulic gradient is as shown in Fig. 11.7.

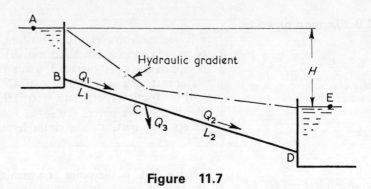

Figure 11.7

11.10 Three-reservoir problem

A reservoir, surface level 60 m above datum, supplies a junction box through a 300 mm pipe, 1500 m long. From the junction box two 300 mm pipes, each 1500 m long, feed respectively into two reservoirs whose surface levels are 30 m and 15 m above datum, f for all pipes being 0·01. What will be the quantity entering each reservoir?

Solution. In Fig. 11.8 the velocities v_1, v_2 and v_3 are unknown. Three equations are therefore required and these are Bernoulli's equation applied to A and B, Bernoulli's equation applied to A and C, and the continuity of flow equation.

Apply Bernoulli's equation to flow from A to B:

$$H_A = H_B + \frac{4fL_1}{d_1}\frac{v_1^2}{2g} + \frac{4fL_2}{d_2}\frac{v_2^2}{2g}$$

$$60 = 30 + \frac{4 \times 0{\cdot}01 \times 1500}{0{\cdot}3}\frac{v_1^2}{2g} + \frac{4 \times 0{\cdot}01 \times 1500}{0{\cdot}3}\frac{v_2^2}{2g}$$

$$10{\cdot}2v_1^2 + 10{\cdot}2v_2^2 = 30 \qquad . \qquad . \qquad . \quad (1)$$

Apply Bernoulli's equation to flow from A to C:

$$H_A = H_C + \frac{4fL_1}{d_1}\frac{v_1^2}{2g} + \frac{4fL_3}{d_3}\frac{v_3^2}{2g}$$

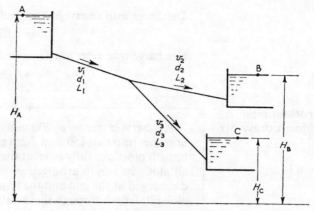

Figure 11.8

$$60 = 15 + \frac{4 \times 0.01 \times 1500}{0.3} \frac{v_1{}^2}{2g} + \frac{4 \times 0.01 \times 1500}{0.3} \frac{v_2{}^2}{2g}$$

$$10.2v_1{}^2 + 10.2v_3{}^2 = 45 \qquad . \qquad . \qquad . \quad (2)$$

For continuity of flow

$$\tfrac{1}{4}\pi d_1{}^2 v_1 = \tfrac{1}{4}\pi d_2{}^2 v_2 + \tfrac{1}{4}\pi d_3{}^2 v_3$$

or, since $d_1 = d_2 = d_3$,

$$v_1 = v_2 + v_3 \qquad . \qquad . \qquad . \quad (3)$$

From equation (1) $\qquad v_2 = \sqrt{(2.94 - v_1{}^2)}$

From equation (2) $\qquad v_3 = \sqrt{(4.42 - v_1{}^2)}$

Substituting in equation (3):

$$v_1 - \sqrt{(2.94 - v_1{}^2)} - \sqrt{(4.42 - v_1{}^2)} = 0$$

This equation is solved by successive approximation, or graphically.

If the square roots are to be real, v_1 cannot exceed the value given by $v_1{}^2 = 2.94$ or $v_1 = 1.71$.

Call the left-hand side of the equation $f(v_1)$ and, choosing a value of v_1 less than 1.71, calculate $f(v_1)$.

If $v_1 = 1.6\,\text{m/s}$

$$f(v_1) = 1.6 - 0.6 - 1.36 = -0.36$$

If $v_1 = 1.7\,\text{m/s}$

$$f(v_1) = 1.7 - 0.1 - 1.22 = +0.38$$

Clearly the required value of v_1 lies about midway between 1.6 and 1.7 m/s. A third value can be taken and the three points plotted, or further approximations can be made to obtain the result.

$$v_1 = 1.663\,\text{m/s}$$

$$v_2 = \sqrt{(2.94 - 2.78)} = 0.4\,\text{m/s}$$

$$v_3 = \sqrt{(4.42 - 2.78)} = 1.28\,\text{m/s}$$

$$\text{Discharge into reservoir B} = \tfrac{1}{4}\pi d_2^2 v_2 = \tfrac{1}{4}\pi \times (0\cdot3)^2 \times 0\cdot4$$
$$= 0\cdot028\,\text{m}^3/\text{s} = \textbf{28·3dm}^3/\textbf{s}$$
$$\text{Discharge into reservoir C} = \tfrac{1}{4}\pi d_3^2 v_3 = \tfrac{1}{4}\pi \times (0\cdot3)^2 \times 1\cdot28$$
$$= 0\cdot0907\,\text{m}^3/\text{s} = \textbf{90·7dm}^3/\textbf{s}$$

11.11 Forked pipe with uniform draw off

A reservoir feeds a 200 mm diam pipe 300 m long which branches into two 150 mm diam pipes each 150 m long. Both the branch pipes are fully open at their ends. One branch has outlets all along its length arranged so that half the water entering it is discharged at the end and the remainder is discharged uniformly along its length through these outlets. The far ends of the branch pipes are at the same level, which is 15 m below that of the reservoir. Calculate the discharge from each branch pipe. Disregard all losses except friction and take $f = 0\cdot006$ for all pipes.

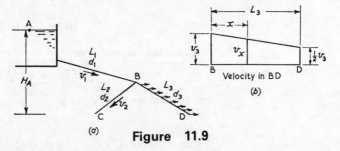

Figure   11.9

Solution. Fig. 11.9(*a*) shows the arrangement of the pipes and Fig. 11.9(*b*) shows the variation of velocity in BD due to the uniform discharge along its length. At a distance x from B

$$v_x = v_3 - \frac{v_3 x}{2L}$$

where $v_3 =$ velocity in BD at B.

For a small length of pipe δx the loss of head is

$$\delta h_f = \frac{4f\delta x}{d_3 2g}\left(v_3 - \frac{v_3 x}{2L}\right)^2$$

Integrating from $x = 0$ to $x = L_3$

$$\text{Loss of head in BD} = \frac{4fv_3^2}{d_3 2g}\int_0^{L_3}\left(1 - \frac{x}{L_3} + \frac{x^2}{4L_3}\right)dx$$

$$= \frac{7}{11}\frac{4fL_3}{d_3}\frac{v_3^2}{2g}$$

For flow from A to D, by Bernoulli's theorem

$$H_A = \frac{4fL_1}{d_1}\frac{v_1^2}{2g} + \frac{7}{12}\frac{4fL_3}{d_3}\frac{v_3^2}{2g}$$

taking friction only into account.

$$15 = \frac{4 \times 0.006 \times 300}{0.2 \times 2g} v_1^2 + \frac{7}{12} \times \frac{4 \times 0.006 \times 150}{0.15 \times 2g} v_3^2$$

$$= 1.836 v_1^2 + 0.714 v_3^2$$

$$v_3 = \sqrt{(21 - 2.57 v_1^2)} \qquad . \qquad . \qquad . \qquad . \qquad (1)$$

For flow from A to C by Bernoulli's theorem

$$H_A = \frac{4fL_1}{d_1} \frac{v_1^2}{2g} + \frac{4fL_2}{d_2} \frac{v_2^2}{2g}$$

$$15 = 1.836 v_1^2 + \frac{4 \times 0.006 \times 150}{0.15 \times 2g} v_2^2$$

$$15 = 1.836 v_1^2 + 1.225 v_2^2$$

$$v_2 = \sqrt{(12.25 - 1.5 v_1^2)} \qquad . \qquad . \qquad . \qquad . \qquad (2)$$

For continuity of flow,

$$\tfrac{1}{4}\pi d_1^2 v_1 = \tfrac{1}{4}\pi d_2^2 v_2 + \tfrac{1}{4}\pi d_3^2 v_3$$

$$(0.2)^2 v_1 = (0.15)^2 v_2 + (0.15)^2 v_3$$

or $\qquad v_1 - 0.563 v_2 - 0.563 v_3 = 0$

Substituting from equations (1) and (2)

$$v_1 - 0.563\sqrt{(12.25 - 1.5 v_1^2)} - 0.563\sqrt{(21 - 2.57 v_1^2)} = 0$$

Calling the left-hand side $F(v_1)$ and solving by successive approximation:

If $v_1 = 2.4\,\text{m/s}$

$$F(v_1) = 2.4 - 0.563\sqrt{3.61} - 0.563\sqrt{6.18}$$
$$= 2.4 - 1.07 - 1.39 = -0.06$$

If $v_1 = 2.45\,\text{m/s}$

$$F(v_1) = 2.45 - 0.563\sqrt{3.25} - 0.563\sqrt{5.6}$$
$$= 2.45 - 1.01 - 1.33 = +0.11$$

If $v_1 = 2.42\,\text{m/s}$

$$F(v_1) = 2.42 - 0.563\sqrt{3.45} - 0.563\sqrt{5.9}$$
$$= 2.42 - 1.04 - 1.36 = +0.02$$

By interpolation:

$$v_1 = 2.415 \text{ makes } F(v_1) = 0$$

$$v_2 = \sqrt{(12.25 - 8.74)} = 1.87\,\text{m/s}$$

$$v_3 = \sqrt{(21 - 15)} = 2.44\,\text{m/s}$$

Discharge from BC $= \tfrac{1}{4}\pi d_2^2 v_2 = \tfrac{1}{4}\pi \times (0.15)^2 \times 1.87$
$$= 0.0341\,\text{m}^3/\text{s}$$

Discharge from BD $= \tfrac{1}{4}\pi d_3^2 v_3 = \tfrac{1}{4}\pi \times (0.15)^2 \times 2.44$
$$= 0.0433\,\text{m}^3/\text{s}$$

Problems

1 A pipe, 50 mm in diam, is 6 m long and the velocity of water in the pipe is $2 \cdot 4$ m/s. What loss of head would be saved if the central $1 \cdot 8$ m of pipe were replaced by a 75 mm diam pipe, the changes of section being sudden ($f = 0 \cdot 01$, $C_c = 0 \cdot 62$)?
Answer $0 \cdot 164$ m

2 A pipe of 50 mm diam and 45 m long is connected to a large tank, the entrance to the pipe being 3 m below the surface. The lower end of the pipe which is 6 m below the upper end is joined to a horizontal pipe of 100 mm diam and 75 m long which discharges to atmosphere. Calculate the discharge taking into account the sudden enlargement and entry losses: $f = 0 \cdot 008$ for both pipes.
Answer $4 \cdot 66 \times 10^{-3}$ m/s

3 Determine the discharge through a new 300 mm diam cast-iron pipe 900 m long under a head of 9 m. Take $f = 0 \cdot 005$ ($1 + d/3 \cdot 6$) where d is the diameter of the pipe in metres.
Answer $116 \cdot 6$ dm³/s

4 Two reservoirs, whose surface levels differ by 30 m, are connected by a pipe $0 \cdot 6$ m in diam and 3000 m long. The pipeline crosses a ridge whose summit is 9 m above the level of and 300 m distant from the higher reservoir. Find the minimum depth below the ridge at which the pipe must be laid, if the absolute pressure in the pipe is not to fall below 3 m of water; and calculate the discharge in m³/s; $f = 0 \cdot 0075$.
Answer $0 \cdot 559$ m³/s; 5 m

5 A pump supplies water to a nozzle of 25 mm diam through a pipe 180 m long and 75 mm in diam. The nozzle is at a level 9 m above the pump. Taking the discharge coefficient of the nozzle as $0 \cdot 94$ and the coefficient of friction of the pipe as $0 \cdot 012$, calculate the pressure required at the pump outlet to give a discharge of 8 dm³/s.
Answer 426 kN/m²

6 Two reservoirs $4 \cdot 8$ km apart are connected by a pipeline which consists of 150 mm diam pipe for the first $1 \cdot 6$ km, sloping $5 \cdot 7$ m per km, and 225 mm diam pipe for the remaining distance having a slope of $1 \cdot 9$ m per km. The levels of the water above the pipe openings are 6 m in the upper reservoir and $3 \cdot 6$ m in the lower. Taking $f = 0 \cdot 0075$, calculate the discharge through the pipeline, ignoring losses other than friction. Draw the hydraulic gradient.
Answer $16 \cdot 3$ dm³/s

7 Two tanks are joined by 30 m of 100 mm diam pipe. The difference in level in the tanks is 6 m and the ends of the pipe are 3 m under water. Both pipe ends are sharp and the loss at the entrance may be taken as $0 \cdot 5$ $v^2/2g$. Sketch the hydraulic gradient and find (*a*) the velocity in the pipe, (*b*) the pressure head in the pipe midway along its length: $f = 0 \cdot 01$.
Answer $2 \cdot 94$ m/s; $2 \cdot 68$ m

8 Water discharges from a large tank through a pipe of 50 mm diam and 45 m long which is sharp at entry, after which there is a sudden enlargement to a pipe of 75 mm diam and 30 m long. The point of delivery is 6 m below the water surface in the tank. Determine the discharge in dm³/s. Assume that $f = 0 \cdot 005$ for both pipes.
Answer $4 \cdot 7$ dm³/s

9 Two reservoirs are connected by a pipe whose total length is 360 m. From the upper reservoir the pipe is 300 mm diam for a length of 150 m and the remaining 210 m is 450 mm diam. The velocity of water in the smaller pipe is $1 \cdot 2$ m/s. Find the loss of head (*a*) at the entrance to the pipe, (*b*) at the junction, (*c*) at the exit to the lower reservoir. Also calculate the difference in levels of the water in the two reservoirs. Take $f = 0 \cdot 006$ for the smaller pipe and $0 \cdot 005$ for the larger.
Answer $0 \cdot 0367$ m, $0 \cdot 0227$ m, $0 \cdot 0145$ m $1 \cdot 09$ m

10 Two tanks A and C containing water with a difference of level of 3 m are connected by a pipeline ABC. Pipe AB is 100 mm in diam and starts $1 \cdot 5$ m below the surface in tank A running thence horizontally for 30 m to B. Pipe BC is 150 m in diam and 24 m long. It discharges into tank C $0 \cdot 9$ m below water level. Find the discharge and also the pressure head at B in the smaller pipe. Neglect losses other than pipe friction for which $f = 0 \cdot 004$.

If a meter which gives a loss of head of $\frac{1}{2}v^2/2g$ is introduced into the 100 mm pipe, find the power lost, assuming the same rate of flow.
Answer $0 \cdot 0262$ m³/s; $-1 \cdot 78$ m; $72 \cdot 6$ W

11 $0 \cdot 9$ m³ of oil per minute are pumped through a pipe 600 m long and of 100 mm diam. The oil has an absolute viscosity of $0 \cdot 00525$ kg-m/s and a specific weight of 930 kg/m³. Find the head lost in friction, if the friction coefficient f is given by $f = 0 \cdot 064 \, R^{-0.23}$ where R is the Reynolds number.

If a pump having an efficiency of 65 per cent is used to overcome the friction and also to elevate the oil through 18 m, determine the power required to drive the pump.
Answer 26 m; $9 \cdot 3$ kW

12 A horizontal water main comprises 1500 m of 150 mm diam pipe followed by 900 m of 100 mm diam pipe, the friction coefficient for each pipe being $0 \cdot 007$. All the water is drawn off at a uniform rate per unit of length along the pipe. If the total input to the system is 25 dm³/s, find the total pressure drop along the main, neglecting all losses other than pipe friction. Also draw the hydraulic gradient, taking the pressure head at inlet as 54 m.
Answer $20 \cdot 68$ m

13 A 675 mm water main runs horizontally for 1500 m and then branches into two 450 mm mains each 3000 m long.

In one of these branches the whole of the water entering is drawn off at a uniform rate along the length of the pipe. In the other branch one-half of the quantity entering is drawn off at a uniform rate along the length of the pipe. If $f = 0 \cdot 006$

throughout, calculate the total difference of head between inlet and outlet when the inflow to the system is $0 \cdot 28$ m³/s. Consider only frictional losses and assume atmospheric pressure at the end of each branch.

Answer $4 \cdot 78$ m

14 Two reservoirs having a difference of level of 18 m are connected by a straight pipe, 450 mm in diam and 900 m long, which connects to the upper reservoir $2 \cdot 4$ m below the surface, and to the lower one $1 \cdot 5$ m below the surface. If $f = 0 \cdot 008$, calculate the flow in the pipe and the pressure at a point 300 m from the discharge end.

If a 300 mm main is connected in parallel with the 450 mm main over the last 300 m, what will be the flow from the upper reservoir and the pressure at the branch point?

Answer $0 \cdot 369$ m³/s; $15 \cdot 75$ kN/m²; $0 \cdot 4$ m³/s; $-5 \cdot 9$ kN/m²

15 Two reservoirs A and B discharge through circular pipes, each 600 mm diam and $1 \cdot 6$ km long, to a junction at D. From D the combined discharge is carried to a third reservoir C by a 900 mm diam pipe of negligible length. The surface level at A is 15 m and that at B 9 m above the level of C. Neglecting all losses except pipe friction, find the discharge from each reservoir. Take $f = 0 \cdot 0075$.

Answer $0 \cdot 542$ m³/s, $0 \cdot 420$ m³/s

16 Two pipelines of equal length, 3000 m, are laid in parallel between two reservoirs whose difference of level is 15 m. If their diameters are respectively 300 mm and 600 mm and if the frictional resistance is given by

$$h = \frac{flv^{1 \cdot 8}}{d^{1 \cdot 2}}$$

where $f = 0 \cdot 01$ what will be the total discharge?

Answer $0 \cdot 158$ m³/s

17 Two reservoirs are connected by two pipelines in parallel. Their diameters are 300 mm and 600 mm and their respective lengths are 1500 m and 3000 m. If the value of f for the smaller pipe is $0 \cdot 008$ and for the larger is $0 \cdot 006$, what will be the discharge from the larger when the smaller pipe is delivering $0 \cdot 056$ m³/s.

Answer $0 \cdot 256$ m³/s

18 Two sharp-ended pipes of 50 mm and 100 mm diam, each 30 m long, are connected in parallel between two reservoirs whose difference of level is $7 \cdot 5$ m. Find (*a*) the flow in dm³/s for each pipe and draw the corresponding hydraulic gradients, (*b*) the diameter of a single pipe 30 m long which will give the same flow as the two actual pipes. Assume that the entrance loss is $0 \cdot 5v^2/2g$ and take $f = 0 \cdot 008$ in each case.

Answer $5 \cdot 24$ dm³/s; $28 \cdot 59$ dm³/s; 107 mm

19 A pipeline conveying water consists of 750 m of an 200 mm diam pipe with another 200 mm pipe running in parallel for the

last 300 m. A gate valve is installed in one of the two branches. If the total flow is reduced by 5 per cent by partially closing the valve, find the loss of head across the valve in terms of the new velocity in that pipe, taking $f = 0 \cdot 007$.
Answer $6 \cdot 43v^2$

20 Water is drawn from a tank, whose surface level is 15 m above the ground, through a 25 mm diam pipe. At A, 18 m along the pipe and 6 m above the ground the pipe branches into two $12 \cdot 5$ mm diam pipes each 12 m in length and terminating at B, $1 \cdot 5$ m above the ground. Calculate the flow from the tank (*a*) when one branch only is open, (*b*) when both branches are open. Take $f = 0 \cdot 01$ and neglect shock losses.
Answer $0 \cdot 312$ dm³/s, $0 \cdot 585$ dm³/s

21 Water flows from a reservoir through a pipe, 150 mm in diam and 180 m long, to a point $13 \cdot 5$ m below the open surface of the reservoir. Here it branches into two pipes, each of 100 mm diam, one of which is 48 m long, discharging to atmosphere at a point 18 m below reservoir level, and the other 60 m long, discharging to atmosphere 24 m below reservoir level. Assuming a friction coefficient of $0 \cdot 008$ and neglecting any loss at the junction, calculate the discharge from each pipe.
Answer $19 \cdot 3$ dm³/s; $25 \cdot 7$ dm³/s

22 Reservoir A at an elevation of 270 m supplies water to reservoirs B and C at levels of, respectively, 180 m and 150 m. From A to D both supplies pass through a common pipe 300 mm in diam and 16 km long. The branch D to B is 225 mm in diam and $9 \cdot 6$ km long and that from D to C is 150 mm in diam and 8 km long. How many cubic decimetres per sec will be delivered to B and C if $f = 0 \cdot 01$?
Answer $29 \cdot 3$, $14 \cdot 9$ dm³/s

23 A reservoir A discharges through a pipe 450 mm in diam and 900 m long which is connected to branched pipes, one 1200 m long leading to reservoir B 36 m below A, and the other 1500 m long leading to C 45 m below A. Calculate the diameters of the branches if they have equal discharges which together equal that of a 450 mm pipe connected directly from A to B. Neglect all losses except friction and assume that *f* is the same for all pipes.
Answer $0 \cdot 342$ m; $0 \cdot 332$ m

24 A 900 mm diam concrete pipe 1500 m long draws water from reservoir A. At its lower end it is joined with a 750 mm diam concrete pipe 1200 m long drawing water from reservoir B, both pipes discharging into a 1200 mm diam concrete pipe 2400 m long. If the discharge end of the 1200 mm pipe is $4 \cdot 5$ m lower than the water surface in A and 6 m lower than the surface in B, determine the discharge. Take $f = 0 \cdot 005$ for all pipes, and ignore all other losses.
Answer $1 \cdot 28$ m³/s

25 Three reservoirs A, B and C are connected by pipelines. The surface levels in B and C are respectively 12 m and 30 m below that in A. From A two pipes, 150 mm and 300 mm in diam and each 300 m long, are laid in parallel to a junction box J, from which a 150 mm pipe 450 m long connects J to B and a 300 mm pipe 900 m long connects J to C. Neglecting all losses other than pipe friction, and taking $f = 0 \cdot 01$ for all the pipes, calculate the flow into each of the reservoirs B and C.

Answer 16 dm³/s; 137 dm³/s

26 Water is discharged from a reservoir through a pipe 150 mm in diam and 120 m long. This pipe divides into two pipes each of 75 mm diam. One is 30 m long and discharges into a second reservoir with water level 12 m below the first, the other is 60 m long and discharges into a third reservoir with water level 24 m below the first. Taking $f = 0 \cdot 01$ for each pipe, find the discharge into each reservoir. Neglect all losses other than pipe friction.

Answer 13·55 dm³/s; 15·35 dm³/s

12

Transmission of power by pipeline

If a pipeline is to transmit power, only a part of the gross head H at the inlet is available to overcome friction. If h_p is the head at the outlet end, h_f the head lost in friction, Q the discharge and w the specific weight of the fluid:

$$\text{Power available at outlet} = wQh_p$$

$$\text{Power supplied at inlet} = wQH$$

$$\text{Efficiency of transmission } \eta = \frac{wQh_p}{wQH} = \frac{h_p}{H}$$

Also $$H = h_p + h_f$$

so that $$h_p = H - h_f$$

$$\text{Efficiency of transmission } \eta = 1 - \frac{h_f}{H}$$

or $$h_f = (1 - \eta)H$$

12.1 Variation of power with discharge

A pipeline of a given length and diameter has a difference of head H between the input and output ends. Show by means of sketch diagrams how the head lost in friction h_f, the power head h_p, the power P and the efficiency of power transmission η vary with the discharge Q.

Solution. Fig. 12.1(a) shows the variation of the head lost in friction h_f with discharge Q.

By the Darcy formula

$$h_f = \frac{4fL}{d}\frac{v^2}{2g} = \frac{fLQ^2}{3d^5} \text{ in SI units}$$

Hence h_f is proportional to Q^2.

Since the power head $h_p = H - h_f$, Fig. 12.1(a) also shows the variation of h_p with Q.

Power transmitted $= wQh_p$ and from Fig. 12.1(a), when $Q = 0$, h_p is a max and power $= 0$; when Q is a max $h_p = 0$ and power $= 0$. At intermediate values of Q both Q and h_p have positive values. The relation between power and discharge will therefore be as in Fig. 12.1(b), rising from zero to maximum and returning to zero.

Efficiency of transmission $\eta = \dfrac{h_p}{H}$

The variation of h_p with Q is shown in Fig. 12.1(a) and for a given value of H the variation of η with Q will be as shown in Fig. 12.1(c).

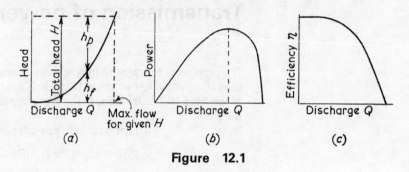

Figure 12.1

12.2 Conditions for maximum power transmission

Determine the conditions for maximum transmission of power through a pipe assuming loss of head by friction only. Calculate the maximum power available at the far end of a hydraulic pipeline 4·8 km long and 200 mm in diam when water at 6900 kN/m² pressure is fed in at the near end. Take the friction coefficient $f = 0·007$.

Solution. If $H =$ supply head and $h_f =$ head lost in friction

Power head at delivery $h_p = H - h_f$

By the Darcy formula

$$h_f = \frac{4fL}{d}\frac{v^2}{2g}$$

where $L =$ length of pipe, $d =$ pipe diam and $v =$ velocity in pipe. Thus,

$$h_p = H - \frac{4fL}{d}\frac{v^2}{2g} = H - cv^2$$

where c is a constant.

Power $P = wavh_p = wav(H - cv^2)$

where $a =$ area of pipe and $w =$ spec. weight of fluid.

For maximum power

$$\frac{dP}{dv} = 0$$

or $\qquad\qquad waH - 3wacv^2 = 0$

$$H = 3cv^2 = 3 \times \frac{4fL}{d} \frac{2g}{v^2} = 3h_f$$

$$h_f = \tfrac{1}{3}H$$

Head lost in friction $= \tfrac{1}{3}$ of head at inlet

If pressure of water at inlet $p = 6900$ kN/m^2

$$H = \frac{p}{w} = \frac{6900 \times 10^3}{9 \cdot 81 \times 10^3} = 704\,\text{m of water}$$

For maximum power:

Head lost in friction $h_f = \tfrac{1}{3}H = 234 \cdot 7$ m of water

Power head $h_p = \tfrac{2}{3}H = 469 \cdot 3$ m of water

Using the Darcy formula,

$$h_f = \frac{4fL}{d} \frac{v^2}{2g}$$

$$v^2 = \frac{2gdh_f}{4fL} = \frac{2g \times 0 \cdot 2 \times 234 \cdot 7}{4 \times 0 \cdot 007 \times 4800}$$

$$= 6 \cdot 87$$

$$v = 2 \cdot 62\,\text{m/s}$$

Maximum power $= wavh_p$

$$= 9 \cdot 81 \times 10^3 \times \tfrac{1}{4}\pi(0 \cdot 2)^2 \times 2 \cdot 62 \times 469 \cdot 3\,\text{W}$$

$$= \mathbf{379\,kW}$$

12.3 Power for a given efficiency

Find the least number of 150 mm diam pipes required to transmit 170 kW to a machine $3 \cdot 2$ km from the power station, if the efficiency of transmission is to be 92 per cent and $f = 0 \cdot 0075$. The feed pressure is 4800 kN/m^2.

Solution. If $p_1 = $ input pressure and $p_2 = $ pressure at machine

$$\text{Input head } H = \frac{p_1}{w}$$

$$\text{Power head } h_p = \frac{p_2}{w}$$

$$\text{Efficiency } \eta = \frac{h_p}{H} = \frac{p_2}{p_1} = 0 \cdot 92$$

$$\therefore \qquad p_2 = 0 \cdot 92 p_1 = 0 \cdot 92 \times 4800 = 4420\,\text{kN/m}^2$$

$$\text{Loss of head in friction } h_f = \frac{(p_1 - p_2)}{w}$$

$$= \frac{4fL}{d} \frac{v^2}{2g}$$

$$v^2 = \frac{2gd}{4fL} \frac{(p_1 - p_2)}{w}$$

$$= \frac{2g \times 0\cdot15 \times (4800 - 4420) \times 10^3}{4 \times 0\cdot0075 \times 3\cdot2 \times 10^3 \times 9\cdot81 \times 10^3} = 1\cdot19$$

$$v = 1\cdot095 \, \text{m/s}$$

Power transmitted by one 150 mm pipe

$$= p_2 a v$$

$$= 4420 \times 10^3 \times \tfrac{1}{4}\pi \times (0\cdot15)^2 \times 1\cdot095 \, \text{W}$$

$$= 85\cdot6 \times 10^3 \, \text{W}$$

$$= 85\cdot6 \, \text{kW}$$

Total power required $= 170 \, \text{kW}$

$$\text{Number of pipes required} = \frac{170}{85\cdot6} = 2$$

12.4 Nozzle size for maximum power

Water from a reservoir flows through a pipeline of length L and diameter d and discharges through a nozzle of diameter d_n.

If the loss of head in the nozzle is $k(v_n^2/2g)$ where k is a constant and v_n the velocity in the nozzle, and f is the friction coefficient for the pipe, show that for maximum power transmission through the pipeline

$$\frac{d_n}{d} = \left[\frac{(1 + k)d}{8fL}\right]^{1/4}$$

A pipe 225 mm in diam and 1500 m long discharges water from a nozzle. The total head measured above the centreline of the nozzle is 270 m, the velocity coefficient for the nozzle is $0\cdot96$ and f for the pipe is $0\cdot006$.

If the overall efficiency of power transmission is 81 per cent, find the power transmitted and the flow in m^3/s.

Solution. If $h_f =$ head lost in friction in the pipe

$$h_p = \text{power head}$$

$$H = h_f + h_p = \text{gross head at inlet}$$

For maximum power transmission in the pipeline, $h_f = \tfrac{1}{3}H$ and $h_p = \tfrac{2}{3}H$ so that $h_f = \tfrac{1}{2}h_p$. If $v =$ velocity of flow in the pipeline

$$h_f = \frac{4fL}{d}\frac{v^2}{2g}$$

The nozzle converts part of h_p into kinetic energy $v_n^2/2g$ and there is a loss $k(v_n^2/2g)$ so that

$$h_p = \frac{v_n^2}{2g} + k\frac{v_n^2}{2g} = (1 + k)\frac{v_n^2}{2g}$$

Since $h_f = \frac{1}{2}h_p$,

$$\frac{4fL}{d}\frac{v^2}{2g} = \frac{(1 + k)}{2}\frac{v_n^2}{2g}$$

$$\frac{v^2}{v_n^2} = \frac{(1 + k)d}{8fL}$$

For continuity of flow

$$\tfrac{1}{4}\pi d^2 v = \tfrac{1}{4}\pi d_n^2 v_n$$

$$\therefore \qquad \frac{d_n}{d} = \left(\frac{v}{v_n}\right)^{1/2} = \left\{\frac{(1 + k)d}{8fL}\right\}^{1/4}$$

Velocity head of jet $= \dfrac{v_n^2}{2g} = 0\cdot81 \times H$

$$= 0\cdot81 \times 270 = 219\,\text{m of water}$$

$$v_n = \sqrt{(2g \times 219)} = 65\cdot6\,\text{m/s}$$

If $C_v = $ coefficient of velocity of nozzle,

$$v_n = C_v\sqrt{(2gh_p)}$$

$$h_p = \frac{v_n^2}{2g \times C_v^2} = \frac{219}{(0\cdot96)^2} = 237\,\text{m of water}$$

Loss of head in pipe $h_f = H - h_p = 270 - 237 = 33\,\text{m}$

$$= \frac{4fL}{d}\frac{v^2}{2g}$$

where $v = $ pipe velocity

$$v = \sqrt{\frac{h_f \times 2gd}{4fL}} = \sqrt{\frac{33 \times 2g \times 0\cdot225}{4 \times 0\cdot006 \times 1500}}$$

$$= 2\cdot01\,\text{m/s}$$

Discharge $Q = \tfrac{1}{4}\pi d^2 v = \tfrac{1}{4}\pi(0\cdot225)^2 \times 2\cdot01$

$$= \mathbf{0\cdot08\,m^3/s}$$

Power transmitted $= wQ\dfrac{v_n^2}{2g}$

$$= 9\cdot81 \times 10^3 \times 0\cdot08 \times 219\,\text{W}$$

$$= \mathbf{172\,kW}$$

12.5 Power for a given nozzle size

A pipeline is 1800 m long and 375 mm in diam, and the supply head at inlet is 240 m. A nozzle with an effective diameter of 50 mm is fitted at the discharge end and has a coefficient of velocity of 0·972. If f for the pipe is 0·005, calculate (a) the velocity of the jet, (b) the discharge, (c) the power of the jet.

Solution. (a) Head lost in pipe friction

$$h_f = \frac{4fL}{D}\frac{v^2}{2g}$$

where L = length of pipe, D = pipe diam, v = pipe velocity.
If H = supply head,

$$\text{Head at nozzle} = H - \frac{4fL}{D}\frac{v^2}{2g}$$

$$\text{Jet velocity } V = C_v\sqrt{2g\left(H - \frac{4fL}{D}\frac{v^2}{2g}\right)}$$

$$V^2 = C_v^2 2g\left(H - \frac{4fL}{D}\frac{v^2}{2g}\right) \qquad . \qquad . \qquad . \quad (1)$$

For continuity of flow

$$\tfrac{1}{4}\pi D^2 v = \tfrac{1}{4}\pi d^2 V$$

where d = jet diam

$$v = V\frac{d^2}{D^2}$$

Substituting in equation (1)

$$V^2 = C_v^2 2g\left(H - 4fL\frac{V^2}{2g}\frac{d^4}{D^5}\right)$$

$$= (0{\cdot}972)^2 2g\left(240 - \frac{4\times 0{\cdot}005\times 1800\times V^2\times(0{\cdot}05)^4}{2g\times(0{\cdot}375)^5}\right)$$

$$V^2 = 18{\cdot}56(240 - 0{\cdot}001\,545V^2)$$

$$1{\cdot}028\,6V^2 = 4460$$

$$V^2 = 4350$$

$$\text{Velocity of jet } V = \textbf{66m/s}$$

(b) $\qquad\qquad \text{Discharge} = \tfrac{1}{4}\pi d^2 V$

$$= \tfrac{1}{4}\pi(0{\cdot}05)^2\times 66 = \textbf{0{\cdot}1296m}^3\textbf{/s}$$

(c) Power of jet $= \dfrac{wQV^2}{2g}$

$$= \frac{9{\cdot}81\times 10^3\times 0{\cdot}1296\times 66^2}{2\times 9{\cdot}81} = 282\,500\,\text{W}$$

$$= \textbf{282{\cdot}5kW}$$

12.6 Nozzle and triple pipeline

A single jet Pelton wheel is fed by three equal parallel pipes, each of length L, diameter D and having a coefficient of friction f, connected to the machine through a junction box. Determine the diameter of the theoretically best nozzle through which the joint flow in the pipes should be discharged from the junction box to give maximum power output. Calculate the nozzle diameter and power of the jet if for each pipe $D = 150$ mm, $L = 1350$ m, $f = 0\cdot0075$ and the gross head available is 216 m.

Solution. The arrangement is shown diagrammatically in Fig. 12.2. Let V be the velocity of flow in each pipe.

(a) For maximum power transmission through each pipe

$$\text{Head lost in friction } h_f = \tfrac{1}{3} \text{ input head } H$$

as shown in Example 12.2, and

$$\text{Power head } h_p = \tfrac{2}{3}H$$

Thus

$$h_f = \tfrac{1}{2}h_p \quad . \qquad . \qquad . \qquad . \qquad . \quad (1)$$

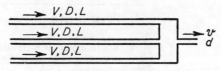

Figure 12.2

Using the Darcy formula

$$h_f = \frac{4fL}{D}\frac{V^2}{2g}$$

Also

$$h_p = \frac{v^2}{2g}$$

where v = jet velocity.

Substituting in equation (1)

$$\frac{4fL}{D}\frac{V^2}{2g} = \frac{1}{2}\cdot\frac{v^2}{2g}$$

$$\frac{8fL}{D} = \frac{v^2}{V^2} \quad . \qquad . \qquad . \qquad . \qquad . \quad (2)$$

For continuity of flow, since there are three supply pipes

$$3(\tfrac{1}{4}\pi)D^2V = \tfrac{1}{4}\pi d^2 v$$

$$\frac{v}{V} = 3\frac{D^2}{d^2}$$

Substituting in equation (2)

$$\frac{8fL}{D} = \frac{9D^4}{d^4}$$

$$d = \sqrt[4]{\frac{9D^5}{8fL}}$$

(b) Putting $D = 150\,\text{mm} = 0 \cdot 15\,\text{m}$, $f = 0 \cdot 0075$ and $L = 1350\,\text{m}$

$$d = \sqrt[4]{\frac{9 \times (0 \cdot 15)^5}{8 \times 0 \cdot 0075 \times 1350}} = 0 \cdot 054\,\text{m}$$

Nozzle diameter $d = \textbf{54\,mm}$

(c) Power head $h_p = \frac{2}{3}H = \frac{2}{3} \times 216 = 144\,\text{m}$

Jet velocity $v = \sqrt{(2gh_p)} = \sqrt{(2g \times 144)}$
$$= 53 \cdot 2\,\text{m/s}$$

$$\text{Power} = wQh_p = w\tfrac{1}{4}\pi d^2 v h_p$$
$$= 9 \cdot 81 \times 10^3 \times \tfrac{1}{4}\pi \times (0 \cdot 054)^2 \times 53 \cdot 2 \times 144$$
$$= 172000\,\text{W} = \textbf{172\,kW}$$

Problems

1 In hydraulic transmission of power, state the losses which occur, and explain how they may be minimized. 75 kW is to be transmitted over a distance of 16 km, the pressure at the inlet of the pipe being 6900 kN/m^2. If the pressure drop per km is to be 43·2 kN/m^2, and if $f = 0 \cdot 006$, find the diameter of the pipe and the efficiency of transmission.
Answer   0·146 m; 90 per cent

2 Water is to be conveyed to a Pelton wheel through a pipe 1200 m long with a fall between open level and the nozzle of 126 m. If the output power is to be 300 kW with a turbine efficiency of 70 per cent, calculate the smallest size pipe which could be employed. Take f as 0·008.
Answer   0·46 m

3 What is the maximum rate at which energy can be transmitted through a 150 mm pipeline 3000 m long supplied with water at 8300 kN/m^2 pressure if $f = 0 \cdot 01$?
Answer   257 kW

4 Water is supplied through a pipe of 50 mm bore 60 m long under a head of 16·5 m and is discharged through a nozzle which has a coefficient of velocity of 0·98. Calculate the bore of the nozzle to give a discharge of 220 dm^3/min if $f = 0 \cdot 006$.
Answer   17·9 mm

5 A pump feeds two hoses each of which is 45 m long and is fitted with a nozzle. Each nozzle has a coefficient of velocity of 0·97 and discharges a 37·5 mm diam jet of water at 24 m/s when the

nozzle is at the same level as the pump. If the power lost in overcoming friction in the hoses is not to exceed 20 per cent of the hydraulic power available at the inlet end of the hoses, calculate (a) the diameter of the hoses, taking $f = 0 \cdot 007$, (b) the power required to drive the pump if its efficiency is 70 per cent and it draws its water supply from a level 3 m below the nozzle.
Answer 98·5 mm; 31·1 kW

6 Determine the diameter of pipe required to supply a turbine developing 3000 kW under a gross head of 150 m if 95 per cent of the head is to be available at the turbine. The efficiency of the turbine is 86 per cent, the pipe is 3000 m long and $C = 66$ in SI units.
Answer 1·3 m

7 Some hydraulic machines are served with water under pressure by a pipe 300 m long, the pressure at the machines being 4140 kN/m². The power available at the machines is 220 kW and the frictional loss of power in the pipes is 88 kW. Find the necessary diameter of pipes if $f = 0 \cdot 03$, and the pressure at the pumping end. What is the maximum power at which the machines could be worked for the same pump pressure?
Answer 137 mm; 5790 kN/m²; 222 kW

8 An hydraulic machine develops 185 kW when the pressure at the machine is 4150 kN/m². The supply pipe is 270 m long and the power lost in friction in the pipe is 7·5 kW. Assuming a friction coefficient of $0 \cdot 0075$ for the pipe, find (a) the diameter of the pipe; and (b) the pressure at the accumulator supplying the pipeline.
Answer 150 mm; 4320 kN/m²

9 The water supply to a Pelton wheel in a power station is 1·27 m³/s under a total head of 285 m and is transmitted through a pipe 720 m long. If the efficiency of transmission is 95 per cent and $f = 0 \cdot 0075$, find the necessary diameter of pipe. What should be the diameter of the jet?
Answer 726 mm; 149 mm

10 The power of a jet driving a Pelton wheel is 1350 kW. The gross head is 270 m and the water is conveyed from the reservoir to the nozzle through a pipe 1500 m long, for which the friction coefficient is $0 \cdot 005$. The velocity coefficient for the nozzle is 0·97. Losses in the pipeline, including the nozzle, are 9 per cent of the gross head. Find (a) the diameter of the pipe, and (b) the diameter of the jet.
Answer 0·62 m; 101 mm

11 A pipeline transmits 260 kW a distance of 2·4 km. If the supply pressure is 3300 kN/m², calculate the number of 150 mm diam pipes required to transmit this power with an efficiency of 92 per cent if $f = 0 \cdot 01$.
Answer 6

12 The output of a multi-cylinder hydraulic motor is required to be 135 kW when its efficiency is 73 per cent. An hydraulic power station developing a pressure of 8250 kN/m² supplies the motor through four 75 mm pipes 3·2 km long. Determine the pressure at the motor, the velocity of flow in the pipes and the efficiency of transmission. Take the friction coefficient f for the pipes as 0·008.

Answer 4417 kN/m²; 2·37 m/s; 53·5 per cent

13 The head of water at the inlet to a pipe 180 m long and 75 mm diam is 30 m and f for the main is 0·01. What diameter of nozzle fitted to the end of the pipe will give maximum power and what power will be available?

Answer 20·2 mm; 1·26 kW

14 The gross head from reservoir level to the four nozzles of a twin-runner Pelton wheel is 300 m. The length of the supply pipeline is 3000 m with $f = 0·005$. The turbine develops 10 500 kW with an efficiency of 85 per cent and the efficiency of power transmission in the pipeline is 90 per cent. If the coefficient of velocity for the nozzles is 0·98, calculate the diameters of (*a*) the pipeline, and (*b*) of each jet.

Answer 1·31 m; 0·146 m

15 A hosepipe of diameter D and length L is fitted with a nozzle of diameter d at the end. If the losses in the nozzle and pipe entry are 10 per cent of the frictional loss in the pipe, show that, for any fixed supply head and pipe, the force of the jet from the nozzle is a maximum when

$$\left(\frac{d}{D}\right) = \left(\frac{D}{4·4fL}\right)^{1/4}$$

where f is the coefficient of friction for the pipe.

A hosepipe 75 mm in diam and 180 m long discharges water from a nozzle fitted to the end of the pipe. The head at entry to the pipe is 39 m, measured above the nozzle, which has a coefficient of velocity of 0·97. The friction coefficient f is 0·009. Find the flow in dm³/s if the useful energy of the jet is 70 per cent of the supply head and all losses except pipe friction and nozzle loss are neglected.

Answer 6·63 dm³/s

16 A turbine is fed by four pipes laid in parallel and connected at the discharge end by a short connecting pipe to the turbine nozzle. Neglecting losses in the connecting pipe, prove that the power of the jet is a maximum when

$$d = \sqrt[4]{\frac{2D^5}{fl}}$$

where d = jet diameter, D = pipe diameter, l = length of each pipe. Determine the jet diameter and velocity of flow and the maximum power delivered by the jet if $D = 600$ mm, $l = 3000$ m, $f = 0·005$. The available head $H = 30$ m.

Answer 0·319 m; 19·85 m/s; 312 kW

17 Show that if η_p is the efficiency of power transmission through a pipeline conveying water to a reaction turbine having efficiency η_t and H is the head from reservoir level to the turbine tailrace, then the power developed by the turbine will be proportional to

$$\eta_t \eta_p \sqrt{(1 - \eta_p)} \, . \, H\sqrt{H}$$

A pipeline 3000 m long conveys water to a turbine which develops 3750 kW with efficiency 90 per cent. The efficiency of power transmission through the pipeline is 91 per cent and the friction coefficient for the pipeline is 0·004. The head from reservoir level to turbine tail-race is 24 m. Find the diameter of the pipeline.

Answer 3·71 m

18 Water is conveyed from a reservoir through a pipe of length 3220 m and at the end of the pipe it is discharged through a nozzle. The surface level of the reservoir is constant and 214·8 m above the nozzle, and the rate of flow is 3·17 m³/s.

Calculate the pipe and nozzle diameters required to enable 85 per cent of the potential energy in the reservoir to be available as kinetic energy in the jet. The coefficient of friction f (wall shear stress/mean dynamic pressure) for the pipe is 0·0075 and all other losses may be neglected.

Answer 1·20 m; 0·26 m

19 A single jet Pelton wheel is to be supplied by a number of pipes in parallel connected to a short common pipe leading to the nozzle. Show that the power delivered to the wheel by the jet is a maximum when the diameter of the nozzle is

$$d = \sqrt{n} \left(\frac{D^5}{8fL} \right)^{1/4}$$

where n = number of pipes, D = diameter of pipes, L = length of pipes, f = friction coefficient. It is assumed that the losses in the short pipe and in the nozzle are small in comparison with the friction losses in the pipes.

Hence find the flow of water and the number of pipes required if $D = 0·6$ m and $L = 2800$ m, when the level of water in the reservoir supplying the Pelton wheel is 300 m above the nozzle and the power of the jet is 10 000 kW. Take $f = 0·005$.

Answer 5·1 m³/s, 4

13

Fluid friction, viscosity and oiled bearings

In a fluid at rest there is, by definition, no shear force between the fluid and a solid boundary or between adjacent layers in the fluid, but when a fluid is in motion there will be differences of velocity between the fluid and the boundary or within the fluid itself and shear stresses are produced which exert a frictional drag. Thus if V is the mean velocity of the fluid relative to a solid boundary,

$$\text{Frictional shear stress at boundary} = kV^n$$

The value of the index n depends on the type of flow and $n = 1$ for viscous (laminar or streamline) flow. For turbulent flow $n = 2$.

In the case of viscous flow the shear stress depends on the coefficient of dynamic viscosity μ of the fluid and the velocity gradient in a direction perpendicular to the direction of flow due to adjacent layers of fluid moving at different velocities.

$$\text{Viscous shear stress } \tau = \mu \times \text{velocity gradient}$$

The coefficient of dynamic viscosity is defined as the shear stress required to move one layer of a fluid past another layer unit distance away in the fluid with unit velocity.

13.1 Froude's experiments

Draw curves showing the nature of the results obtained by Froude in regard to the surface friction of planes, of varying length and of different materials, moving through water. If $f = 2 \cdot 63$ and $n = 1 \cdot 83$, find the power required to overcome the skin resistance of a ship, wetted surface 2200m², when going at 18 knots. One knot = 1·853 km/h.

Solution. Froude's experiments were carried out in 1872. A series of flat boards with different surface finishes were towed on edge at uniform speed through still water and the drag and velocity were recorded. The boards were $\frac{3}{16}$ in. thick, 19 in. deep and varied in length from 1 to 50 ft. The top edge was submerged $1\frac{1}{2}$ in. and the boards had a cut-water at the front for which the resistance was measured separately. It was found that the resistance (1) varied with surface finish, (2) was proportional to v^n where n depends on the nature of the surface, decreases to a limit with increase of length, and was independent of velocity v, (3) increased with length although the resistance per unit area decreases as the length increases tending to a constant value for long lengths.

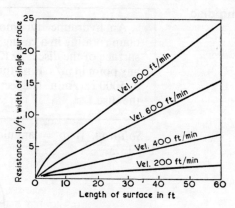

Figure   13.1

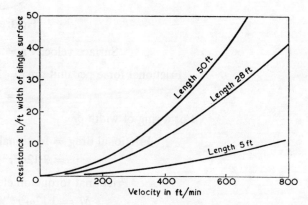

Figure   13.2   (compiled from original experimental data)

If A is the area of the wetted surface, f the frictional resistance per unit area of surface at unit velocity—

$$\text{Total frictional resistance} = fAv^n$$

Fig. 13.1 shows the results obtained for the variation of resistance with length at constant speeds. Fig. 13.2 shows the variation of resistance with velocity for given lengths.

Putting $f = 2·63$, $A = 2200\,\text{m}^2$, $n = 1·83$ and $v = 18\,\text{knots} = 18 \times 1853/3600 = 9·27\,\text{m/s}$

$$\text{Total frictional resistance} = fAv^n$$

$$= 2·63 \times 2200 \times 9·27^{1·83}$$

$$= 339000\,\text{N}$$

$$\text{Power required} = \text{resistance} \times \text{velocity}$$

$$= 339000 \times 9·27 = 3140000\,\text{W}$$

$$= \mathbf{3140\,kW}$$

13.2 Rotating disc

An hydraulic dynamometer consists of a thin disc 200 mm diam rotating in a casing full of water. The frictional drag on the surface of the disc is $3\cdot12v^2\,\text{N/m}^2$ where v is the linear velocity at any point in m/s. Find the power absorbed when the disc is driven at 5000 rev/min. Neglect the effects of the shaft and the edge of the disc.

Solution. If ω = angular velocity of disc, then at any radius r, Fig 13.3,

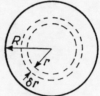

Figure 13.3

$$\text{Surface velocity } v = \omega r$$
$$\text{Frictional force per unit area} = 3\cdot12v^2$$
$$= 3\cdot12\omega^2 r^2\,\text{N/m}^2$$

For a ring of width δr

$$\text{Frictional drag} = \text{frictional force per unit area} \times \text{area}$$
$$= 3\cdot12\omega^2 r^2 \times 2\pi r\delta r\,\text{N}$$

$$\text{Frictional torque on element} = \text{drag} \times \text{radius}$$
$$\delta T = 3\cdot12\omega^2 r^2 \times 2\pi r\delta r \times r\,\text{N-m}$$

Integrating from $r = 0$ to $r = R$, torque on one face is

$$T = 3\cdot12\omega^2 \times 2\pi\int_0^R r^4 dr = \frac{6\cdot24}{5}\pi\omega^2 R^5$$

Since there are two faces,

$$\text{Power absorbed} = 2\omega T = 2\cdot5\pi\omega^3 R^5\,\text{N-m/s}$$

Putting $\omega = (2\pi \times 5000)/60 = 523\,\text{rad/s}$ and $R = 0\cdot1\,\text{m}$

$$\text{Power absorbed} = 2\cdot5\pi(523)^3(0\cdot1)^5\,\text{N-m/s}$$
$$= 11250\,\text{W} = \mathbf{11\cdot25\,kW}$$

13.3. Viscosity

What is meant by (a) the coefficient of dynamic viscosity, (b) the kinematic viscosity of a fluid, (c) the coefficient of dynamic viscosity of a certain oil is $1\cdot5$ cgs units (poises) and its density $0\cdot85$ cgs units? Express the dynamic viscosity in ft lb-wt s units (ft slug s units) and find also the kinematic viscosity of the oil in ft sec units.

Solution. (*a*) When a fluid is in motion shear stresses are developed between layers of the fluid moving with different velocities or between the fluid and a solid boundary. In Fig. 13.4 if a layer at distance y from the boundary moves at velocity v and a layer at $y + dy$ from the boundary at velocity $v + dv$,

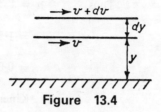

Figure 13.4

Viscous shear stress between layers $= \mu \times$ velocity gradient

or
$$\tau = \mu \, \frac{dy}{dv} \quad . \quad . \quad . \quad . \quad . \quad (1)$$

The coefficient of dynamic viscosity μ is defined as the tangential force per unit area required to move one layer of fluid with unit velocity past another layer in the fluid unit distance away from it.

The units in which μ is measured can be found from equation (1)

$$\mu = \frac{\text{shear stress}}{\text{velocity gradient}} = \frac{\text{force/area}}{\text{velocity/distance}}$$

$$\text{force} = \text{mass} \times \text{acceleration} = \frac{\text{mass} \times \text{length}}{(\text{time})^2}$$

$$\text{area} = (\text{length})^2, \quad \text{velocity} = \frac{\text{length}}{\text{time}}$$

Thus
$$\mu = \frac{\text{mass}}{\text{length} \times \text{time}}$$

(*b*)
$$\text{Kinematic viscosity} = \frac{\text{dynamic viscosity}}{\text{mass density}}$$

The usual symbol is ν, and so

$$\nu = \frac{\mu}{\rho}$$

The units in which kinematic viscosity is measured will be given by

$$\nu = \frac{\text{dynamic viscosity}}{\text{mass density}} = \frac{\text{mass/(length} \times \text{time)}}{\text{mass/(length)}^3} = \frac{(\text{length})^2}{\text{time}}$$

(*c*) The units of dynamic viscosity μ are

$$\frac{\text{length} \times \text{time}}{\text{mass}}$$

To convert from cgs absolute units to ft lb-wt sec (practical) units the units of mass and length must be converted but the unit of time, the second, is the same in both systems.

$$1\,\text{g mass} = \frac{1}{453\cdot6 \times 32\cdot2}\,\text{slugs mass}$$

$$1\,\text{cm} = \frac{1}{30\cdot48}\,\text{ft}$$

Thus $\qquad \mu = 1\cdot5\,\text{cgs abs units}$

$$= 1\cdot5 \times \frac{30\cdot48}{453\cdot6 \times 32\cdot2}\,\text{ft lb-wt sec units}$$

$$= 0\cdot00313\,\text{ft lb-wt sec units or slugs/ft sec}$$

$$\text{Kinematic viscosity } \nu = \frac{\mu}{\rho} = \frac{1\cdot5}{0\cdot85}\,\text{cm}^2/\text{s}$$

and since $\qquad\qquad\qquad 1\,\text{cm} = \frac{1}{30\cdot48}\,\text{ft}$

$$\nu = \frac{1\cdot5}{0\cdot85 \times (30\cdot48)^2} = 0\cdot0019\,\text{ft sec units or ft}^2/\text{s}$$

13.4 Sliding friction

A casting of mass 36 kg slides on a flat plate, which is level and has a velocity of $2\cdot1$ m/s. The area of the base of the casting is 360 cm². Calculate the force required to overcome friction (*a*) if there is no lubricant and the coefficient of solid friction μ_s is $0\cdot2$, (*b*) if there is a lubricating film of oil $0\cdot25$ mm thick of viscosity $\mu = 0\cdot96$ poise.

Solution. (*a*) For solid friction:

$$\text{Frictional force} = \mu_s \times \text{normal force}$$

$$= 0\cdot2 \times 36g = \mathbf{70\cdot5N}$$

(*b*) For fluid friction:

$$\text{Viscous shear stress } \tau = \mu \times \text{velocity gradient}$$

$$\text{Velocity of casting } v \text{ relative to plate} = 2\cdot1\,\text{m/s}$$

$$\text{Thickness of film } t = 0\cdot00025\,\text{m}$$

$$\text{Velocity gradient} = \frac{2\cdot1}{0\cdot00025}\,\text{m/s per m}$$

$$\text{Viscosity} = 0\cdot96\,\text{poise} = 0\cdot096\,\text{kg/m-sec}$$

$$\text{Viscous shear stress } \tau = \mu\frac{v}{t} = \frac{0\cdot096 \times 2\cdot1}{0\cdot00025} = 807\,\text{N/m}^2$$

$$\text{Frictional force} = \tau \times \text{area of base}$$

$$= 807 \times 0\cdot036 = \mathbf{29N}$$

13.5 Journal bearing

A shaft of diameter 74·90 mm rotates in a bearing of diameter 75·03 mm and of length 75 mm. The annular space between the shaft and the bearing is filled with oil having a coefficient of viscosity 0·096 kg/m-s. Determine the power used in overcoming viscous resistance in this bearing at 1400 rev/min.

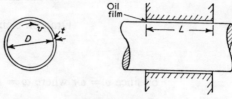

Figure   13.5

Solution. It is assumed that shaft and bearing are concentric (Fig. 13.5) and that the velocity gradient in the oil film is uniform. Since bearing is at rest:

$$\text{Velocity gradient} = \frac{v}{t}$$

$$\text{Viscous shear stress} = \mu \times \text{velocity gradient} = \mu\frac{v}{t}$$

$$\text{Viscous force on shaft} = \text{shear stress} \times \text{surface area}$$

$$= \mu\frac{v}{t} \times \pi DL$$

$$\text{Power absorbed by friction} = \text{Viscous force} \times \text{peripheral velocity}$$

$$= \frac{\mu v^2 \pi DL}{t}$$

Putting $\mu = 0{\cdot}096\,\text{kg/m-s}$, $D = 0{\cdot}0749\,\text{m}$, $L = 75\,\text{mm} = 0{\cdot}075\,\text{m}$

$$v = \frac{\pi DN}{60} = \frac{1400 \times 0{\cdot}0749 \times \pi}{60} = 5{\cdot}5\,\text{m/s}$$

$$t = \frac{0{\cdot}07503 - 0{\cdot}07490}{2} = 0{\cdot}000065\,\text{m}$$

$$\text{Power absorbed} = \frac{0{\cdot}096 \times (5{\cdot}5)^2 \times \pi \times 0{\cdot}0749 \times 0{\cdot}075}{0{\cdot}000065}\,\text{W}$$

$$= \mathbf{792\,W}$$

13.6 Footstep bearing

The thrust at the lower end of a vertical shaft is taken by a flat disc 100 mm in diam separated from a flat housing by an oil film 0·25 mm thick. If the shaft rotates at 1000 rev/min, and the viscosity of the oil is 1·3 poise, calculate the power absorbed by fluid friction.

Solution. The surface velocity of the disc and therefore the velocity gradient and viscous shear stress in the oil will vary with radius. Let the velocity be v at radius r (Fig. 13.3) and consider an annular element of width δr.

$$\text{Average velocity gradient at radius } r = \frac{v}{t}$$

$$\text{Viscous shear stress at radius } r = \mu \frac{v}{t}$$

$$\text{Drag on element} = \text{shear stress} \times \text{area}$$

$$= \mu \frac{v}{t} \times 2\pi r \delta r$$

or since $v = \omega r$ where $\omega =$ angular velocity of shaft,

$$\text{Drag on element} = \frac{2\pi\mu\omega}{t} r^2 \delta r$$

$$\text{Viscous torque on element} = \text{drag} \times \text{radius}$$

$$= \frac{2\pi\mu\omega r^3 \delta r}{t}$$

$$\text{Power absorbed by element} = \frac{2\pi\mu\omega^2 r^3 \delta r}{t}$$

Integrating from $r = 0$ to $r = R$

$$\text{Total power absorbed} = \frac{\pi\eta\omega^2 R^4}{2t}$$

Now

$$\mu = 1\cdot3 \text{ poise} = 0\cdot13 \text{ kg/m-s}$$

$$\omega = \frac{2\pi \times 1000}{60} = 104\cdot8 \text{ rad/s}$$

$$t = 0\cdot25 \text{ mm} = 0\cdot00025 \text{ m}, \quad R = 0\cdot05 \text{ m}$$

$$\text{Power absorbed in friction} = \frac{\pi \times 0\cdot13 \times (104\cdot8)^2 \times (0\cdot05)^4}{2 \times 0\cdot00025}$$

$$= \mathbf{56\,W}$$

13.7 Michell bearing

Fig 13.6 shows the essential parts of a Michell bearing. Prove that the friction drag of a pallet on the thrust plate is given by

$$\frac{\mu v}{\phi}\left(4 \log \frac{H}{h} - 6\frac{H-h}{H+h}\right)$$

per unit width normal to the direction of sliding, where $\mu =$ viscosity of oil, $v =$ surface speed of thrust plate, $\phi =$ angle of inclination of pallets, H, $h =$ maximum and minimum oil film thickness.

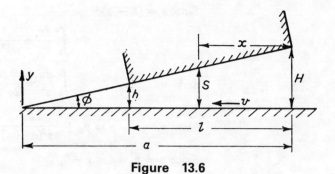

Figure 13.6

Solution.

$$\text{Viscous shear stress } \tau = -\mu \frac{du}{dy}$$

where u is the velocity at a distance y from the thrust plate.

Also if p is the pressure at x

$$\frac{dp}{dx} = \mu \frac{\partial^2 u}{\partial y^2}$$

so that

$$\mu \frac{\partial u}{\partial y} = \frac{dp}{dx} y + A$$

and

$$\mu u = \frac{1}{2} \frac{dp}{dx} y^2 + Ay + B$$

Since $u = v$ when $y = 0$, then $B = \mu v$ and since $u = 0$ when $y = s$,

$$A = -\left(\frac{\mu v}{s} + \frac{dp}{dx} \cdot \frac{s}{2} \right)$$

Thus

$$u = \frac{dp}{dx} \frac{y^2 - sy}{2\mu} + \frac{v}{s}(s - y)$$

$$\text{Shear stress at thrust plate} = \tau_0 = -\mu \left(\frac{\partial u}{\partial y} \right)_0 = -A$$

$$\tau_0 = \frac{\mu v}{s} + \frac{dp}{dx} \frac{s}{2}$$

Also

$$\text{Rate of flow of oil } Q = \int_0^s u\, dy$$

$$Q = \frac{vs}{2} - \frac{dp}{dx} \frac{s^3}{12\mu}$$

Thus

$$\frac{dp}{dx} = 12\mu \left(\frac{v}{2s^2} - \frac{Q}{s^3} \right)$$

and

$$p = p_0 + 12\mu \left\{ \frac{v}{2} \int_0^z \frac{dx}{s^2} - Q \int_0^z \frac{dx}{s^3} \right\}$$

Since $s = (a - x)\phi$

$$\int_0^z \frac{dx}{s^2} = \frac{x}{a\phi^2(a - x)}$$

and

$$\int_0^z \frac{dx}{s^3} = \frac{2ax - x^2}{2\phi^3 a^2(a - x)^2}$$

$$p = p_0 + \frac{6\mu x}{a\phi^2(a - x)}\left(v - \frac{Q(2a - x)}{a\phi(a - x)}\right)$$

Since $p = p_0$ when $x = l$,

$$Q = \frac{va\phi(a - l)}{2a - l}$$

Thus

$$\frac{dp}{dx} = 12\mu\left\{\frac{v}{2s^2} - \frac{va\phi(a - l)}{s^3(2a - l)}\right\}$$

and

$$\tau_0 = \frac{\mu v}{s} + \frac{dp}{dx} \cdot \frac{s}{2}$$

$$= \frac{\mu v}{s} + \frac{12\mu s}{2}\left\{\frac{v}{2s^2} - \frac{va\phi(a - l)}{s^3(2a - l)}\right\}$$

$$= \frac{\mu v}{s}\left\{4 - \frac{6a\phi}{s}\frac{(a - l)}{(2a - l)}\right\}$$

$$= \frac{\mu}{\phi}\left\{\frac{4}{(a - x)} - \frac{6Hh}{(H + h)\phi} \cdot \frac{1}{(a - x)^2}\right\}$$

Shear force per unit width $= \displaystyle\int_0^{a-l} \tau_0 dx$

$$F = \frac{\mu v}{\phi}\int_0^{a-l}\left\{4\frac{dx}{(a - x)} - \frac{6Hh}{(H + h)\phi}\frac{dx}{(a - x)^2}\right\}$$

$$= \frac{\mu v}{\phi}\int_{(a-x)=h/\phi}^{(a-x)=H/\phi}\left\{4\frac{d(a - x)}{(a - x)} - \frac{6Hh}{(H + h)\phi}\frac{d(a - x)}{(a - x)^2}\right\}$$

$$= \frac{\mu v}{\phi}\left\{4\log\frac{H}{h} - \frac{6(H - h)}{(H + h)}\right\}$$

Problems

1 The resistance to motion in the direction of its plane of a thin flat body through water is proportional to v^2 and at 3 m/s, is 24 N/m². Determine the power required to rotate a submerged disc of the same material, 600 mm in diam, at 1200 rev/min.
Answer 32·6 kW

2 A thin flat disc enclosed in a casing containing water is to be used as an hydraulic dynamometer for absorbing and measuring the output from a petrol engine running at 1800 rev/min. Experiments on a similar type of surface show that its frictional resistance per m³ is equal to $2 \cdot 67v^2$N, where v is the velocity in m/s. What diameter of disc will be necessary if the engine develops 37 kW?

Answer 484 mm

3 A hydraulic dynamometer for absorbing the output of an engine on test consists of three thin discs of 600 mm diam mounted on a shaft of 50 mm diam and immersed in water in a casing. If the frictional drag is 12 N/m² at a point where the speed is 3 m/s and if friction is proportional to v^2, what will be the torque on the shaft when the speed is 1200 rev/min?

Answer 386 N-m

4 A circular disc 600 mm diam is immersed in water and rotates at 1200 rev/min. The frictional drag of the surfaces is $1 \cdot 92$ N/m² at a point where the velocity is 3 m/s. What will be the torque required to maintain the rotation if friction is proportional to the square of the velocity?

Answer $20 \cdot 5$ N-m

5 The resistance due to surface friction of a flat plate 1 m² in area when moving in its own plane in water was found to be $34 \cdot 5$ N at 3 m/s and $124 \cdot 5$ N at 6 m/s. Estimate the power absorbed by a thin disc 500 mm in diam having a similar surface when rotated in water at 1500 rev/min. The disc is attached to a shaft of 25 mm diam which extends through it.

Answer $13 \cdot 25$ kW

6 A thin disc of 300 mm diam is mounted on a 20 mm diam shaft and rotates in a closed casing full of water. The surface friction per unit area is proportional to $v^{1 \cdot 8}$ and has a value of $0 \cdot 17$ N/m² when v is $0 \cdot 3$ m/s. Calculate from first principles the power absorbed when the disc rotates at 2000 rev/min.

Answer $1 \cdot 36$ kW

7 A hydraulic dynamometer consists of three parallel discs keyed to the power shaft and rotating between a series of similar stationary discs which form part of a casing filled with water. The outer diameter of the discs is $0 \cdot 6$ m and the inner diameter $0 \cdot 4$ m, the discs being wetted on both sides. The frictional resistance to the motion of the discs in water per m² of wetted area is given by $R = 2 \cdot 145v^{1 \cdot 85}$ in which v is in m/s and R is in newtons.

If 60 kW of power is to be absorbed, calculate the rotational speed in rpm at which the dynamometer must be run.

Answer 3090 rev/min

8 Assuming that $R = fAv^2$ to be the law of friction between a flow and a surface, find an expression for the work lost per second when a disc of radius r is rotated in water with a circumferential velocity v.

If the disc is surrounded by a forced vortex of double its diameter, compare the losses due to the friction of the vortex on the flat sides of the vortex chamber with the loss due to the friction on the above-mentioned disc.

Answer $\frac{4}{3}\pi f v^3 r^2$; 31 to 1

9 The viscosity of water at 0 deg C is 0·01793 poises. Express this value in SI units. What will be the viscosity of water in SI units at 60 deg C? What will be the kinematic viscosity at this temperature in SI units?

Answer 0·001793 kg/ms; 0·47 × 10⁻³ kg/ms; 0·47 × 10⁻⁶m²/s

10 Define "coefficient of viscosity" and derive an expression for its value in terms of Mass, Length and Time. Find the ratio of the coefficient expressed in terms of pounds (weight) force, feet and seconds units compared with the coefficient in SI units. Take 1 lb = 453·6 g, 1 ft = 30·48 cm.

A smooth cylinder 50·1 mm inside diam and 100 mm long is placed with its axis vertical. If the clearance space is entirely filled with oil of viscosity 2·5 poises, calculate the force to push a shaft of 50 mm diam through the cylinder with a velocity of 0·6 m/s.

Answer 47·1 N

11 A cylindrical plug of 100 mm diam and 150 mm long slides concentrically in a stationary cylinder of 100·1 mm internal diam. The clearance space is filled with oil of viscosity 0·175 kg/m-s (N-s/m²). Find from first principles (*a*) the force required to slide the plug along the clyinder at a speed of 3 m/s against the viscous resistance of the oil, (*b*) the work done in one stroke of 0·6 m being half a cycle of SHM with maximum velocity 3 m/s. State what assumptions, if any, are made.

Answer 496 N; 233 J

12 A shaft, 75 mm in diam, revolves concentrically in a bearing 150 mm long. The radial clearance is 1 mm and the speed is 3000 rev/min. The viscosity of the oil is 3 poises. Find the resisting torque due to viscosity and the power absorbed in overcoming it.

Answer 4·684 N-m; 1·47 kW

13 A shaft of 75 mm diam revolves concentrically in a fixed tube of diameter 75·5 mm and length 300 mm. The annular space is full of oil and it is found that a torque of 1 N-m is required to drive the shaft at 2400 rev/min. What is the coefficient of viscosity of the oil? What would be the critical velocity of this oil when flowing in a pipe of 75 mm diam if the critical Reynolds number is taken as 2300 and the density of the oil is 960 kg/m³?

Answer 0·01kg/m-s; 0·319 m/s

14 A shaft, 150 mm in diam, turns concentrically in a sleeve 150·15 mm in diam and 225 mm long. The clearance space is filled with oil. The power required to turn the shaft at 1000 rev/min is 2·2 kW. Find the coefficient of viscosity of the oil, stating clearly the units employed.

Answer 0·0252 kg/m-s

15 A shaft, 100 mm in diam, revolves concentrically in a bearing 150 mm long. The radial clearance is $1 \cdot 25$ mm and the speed is 3000 rev/min. The viscosity of the oil is $2 \cdot 8$ poises. Find the resisting torque due to viscosity and the power absorbed in overcoming it.

Answer $8 \cdot 29$ N-m; $2 \cdot 6$ kW

16 A 150 mm diam shaft runs in a 300 mm long bearing. If an oil of viscosity $1 \cdot 54$ poises and $0 \cdot 0625$ mm thick fills the space between the shaft and the bearing, find the power absorbed when the shaft rotates at 300 rev/min.

Answer $1 \cdot 54$ kW

17 The thrust of a shaft is taken by a collar bearing. A thin film of oil of uniform thickness is maintained between the surfaces of the collars and the bearing pads. The bearing surfaces are three annular pads of inside diam 100 mm and outside diam 150 mm. The thickness of the oil film is $0 \cdot 25$ mm and the coefficient of viscosity is $0 \cdot 85$ poise. Find the power lost in overcoming the viscous torque in the bearing when the shaft rotates at 300 rev/min.

Answer 40 W

18 The thrust of a shaft is taken by a collar bearing fitted with a forced lubrication system which maintains a film of oil of constant thickness $0 \cdot 3$ mm between the surface of the collar and the surface of the bearings. The outer and inner diams of the collar are 160 mm and 120 mm respectively. The coefficient of viscosity of the oil is $0 \cdot 12$ kg/m-s. Calculate the power lost in friction of the bearing when the shaft rotates at 500 rev/min.

Answer $0 \cdot 048$ kW

19 Two plane surfaces, one being fixed and the other moving with velocity U, are separated by a convergent film of lubricant drawn between them by the moving surface.

Assuming that the flow is viscous and two-dimensional and the film very thin so that there is no pressure change across it, i.e. $dP/dy = 0$, show that the rate of change of pressure in the film in the direction of flow is

$$\frac{dP}{dx} = \frac{6\mu U}{h^3} (h - h_0),$$

where h is the film thickness at the point considered, h_0 is the thickness where the pressure P has maximum value and μ is the viscosity of the lubricant.

Also show that, if L is the length of the fixed surface, the tractive force per unit transverse width required to maintain the steady motion of the moving surface is

$$R = \int_0^L \left(\frac{dP}{dx} \cdot \frac{h}{2} + \frac{\mu U}{h} \right) dx$$

14

Flow past a solid boundary

When a fluid flows along a solid boundary a frictional drag is exerted by the boundary causing the velocity of the fluid immediately adjoining the boundary to be the same as that of the boundary. As we move away perpendicularly from the boundary the velocity of the fluid will increase until it reaches that of the fluid in the free stream. The region in which this change of velocity occurs is referred to as the *boundary layer* and its thickness for practical purposes can be defined as the distance from the boundary at which the fluid velocity is 99 per cent of the free stream velocity.

14.1 Flow along a solid boundary, laminar and turbulent boundary layer

(*a*) Describe the formation of a boundary layer flow formed on a solid, smooth, flat surface which is aligned in the direction of flow.

(*b*) Use diagrams to describe the features of a laminar and a turbulent boundary layer and discuss the factors which affect the transition from laminar to turbulent flow.

Solution (*a*) In general all fluids wet all solid surfaces with which they come into contact. A thin layer of fluid remains firmly attached to the surface even when there is relative movement between the solid surface and the main body of the fluid. A *no slip* condition is said to exist at the solid/fluid interface.

For flow over and past a smooth flat stationary surface, the fluid velocity relative to the surface would vary from zero at the solid surface up to the free stream velocity u_0 at a certain distance away. The region in which this change of fluid velocity occurs is called the *boundary layer*.

Within this boundary layer shear stresses will be caused by velocity gradients between adjoining layers of fluid. For laminar flow

$$\text{Shear stress } \tau = \mu \, \frac{du}{dy}$$

where μ is the dynamic viscosity of the fluid and u is the velocity at a distance y measured perpendicularly from the solid surface.

The factors leading to the formation of a boundary layer are therefore the no slip condition at the solid/fluid interface and the shear stresses due to the resulting velocity gradient.

(*b*) A boundary layer growing on a thin flat plate with a sharp leading edge will begin as a laminar boundary layer as shown in Fig. 14.1.

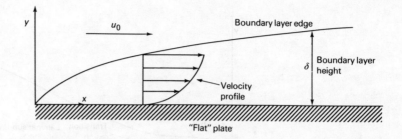

Figure 14.1

At any distance x from the leading edge the boundary layer height δ and hence the position of the boundary layer edge is such that at a distance y from the surface

$$u < u_0 \quad \text{if} \quad y < \delta$$
$$u = u_0 \quad \text{if} \quad y \geqslant \delta$$

Since in reality u will not reach u_0 until $y = \infty$ the position of the boundary layer edge is often taken as the value of y for which $u = 0 \cdot 99\, u_0$.

For further analysis of the boundary layer it is necessary to know the relationship between u and y. Blasius proposed the assumption that the velocity profile within the boundary layer at any distance x from the leading edge was independent of x and the ratio u/u_0 was only a function of y/δ. Common assumptions are that $u/u_0 = y/\delta$ or

$$\frac{u}{u_0} = \sin\left(\frac{\pi}{2} \cdot y/\delta\right) \quad \text{or} \quad \frac{u}{u_0} = 2\frac{y}{\delta} - \frac{y^2}{\delta^2}$$

All these approximations show that the boundary layer height δ varies as $x^{1/2}$ (see Example 14.3).

The shear stress τ in the laminar boundary layer is given by

$$\tau = \mu\,\frac{du}{dy}$$

and varies from zero at the boundary layer edge to a maximum at the solid surface of

$$\tau_w = \mu\left(\frac{du}{dy}\right)_{y=0}$$

After a certain growth distance from the leading edge the laminar boundary layer reaches a *transition point* and flow in the boundary layer becomes turbulent, Fig. 14.2. The boundary layer edge is defined in the same way as for the laminar boundary layer but the velocity profile now takes the form of $u/u_0 = (y/\delta)^{1/n}$ where n is 7 for a fully developed turbulent boundary layer. The velocity profiles are independent of the distance x from the leading edge in the same way as for laminar boundary layer.

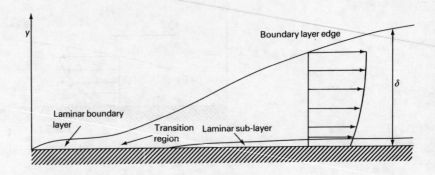

Figure 14.2

Near the wall turbulent fluctuations are damped out and a *laminar sublayer* is formed.

Due to the turbulent motion in the boundary layer fluid is entrained from the main stream at a greater rate than is the case for a laminar boundary layer. High velocity fluid finds its way close to the solid surface rapidly after entrainment. This causes the characteristic "full" velocity profile for turbulent boundary layers.

Shear stress varies from a maximum at the solid boundary to zero at the boundary layer edge. These shear stresses are due to viscous effects near the solid boundary and to turbulent eddy effects in the region between the laminar sublayer and the boundary layer edge. An empirical expression is used for the shear stress at the boundary wall in turbulent boundary layers:

$$\tau_w = 0 \cdot 0225 \, \rho_0 u_0^2 \left(\frac{\mu_0}{\rho_0 u_0 \delta} \right)^{1/4}$$

This has been derived from measurements on turbulent flow in pipes.

Transition from a laminar to a turbulent boundary layer occurs when the structure of the laminar boundary layer becomes unstable at a distance x_T from the leading edge. From experiments it has been found that the Reynolds number at transition $\text{Re}_{x_T} = \rho_0 u_0 x_T / \mu_0$ varies from 10^5 up to a maximum of 2×10^6. For a high level of free stream turbulence or where the boundary surface is rough values nearer to 10^5 should be assumed while for a laminar free stream and a smooth flat boundary transition will not be delayed beyond 2×10^6 in engineering applications.

(*a*) Explain the significance of the boundary layer displacement thickness $\delta*$ and the momentum thickness θ and express them as integrals of the boundary layer velocity profiles on a smooth flat plate.

(*b*) Calculate the ratio $\delta*/\delta$ for a laminar boundary layer with a velocity profile given by

$$\frac{u}{u_0} = 2\frac{y}{\delta} - \frac{y^2}{\delta^2}$$

(*c*) Calculate the ratio θ/δ for a turbulent boundary layer velocity profile given by

$$\frac{u}{u_0} = \left(\frac{y}{\delta}\right)^{1/7}$$

Solution (*a*) The *displacement thickness* $\delta*$ of the boundary layer is the amount by which an ideal inviscid flow would have to be displaced above the solid boundary to satisfy the continuity of mass equation for the combined flow through the boundary layer and the free stream. Thus in Fig. 14.3 (a) there is a real flow with a free stream velocity u_0 beyond the boundary layer and a flow within the boundary layer varying in velocity from u_0 to zero. For the purposes of analysis this can be regarded as equivalent to an inviscid flow only slipping over the displacement surface as shown in Fig. 14.3 (b). The displacement thickness can be determined by applying the continuity of mass equation to the real and inviscid flows as follows.

In Fig. 14.3 (a) the mass flow rate $\dot{m}$ through the space between the solid surface and some arbitrary point Y is, for unit width perpendicular to the diagram

$$\dot{m} = \rho_0 u_0 (Y - \delta) + \int_{y=0}^{y=\delta} \rho_0 u \, dy$$

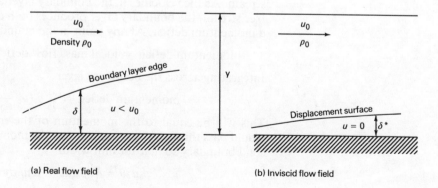

(a) Real flow field   (b) Inviscid flow field

Figure 14.3

In Fig. 14.3 (b) the mass flow rate through the space between the δ^* displacement surface and the same point Y will be, for unit width,

$$\dot{m} = \rho_0 u_0 (Y - \delta^*)$$

For these mass flows to be the same

$$\rho_0 u_0 \delta^* = \rho_0 u_0 \delta - \int_0^\delta \rho_0 u \, dy$$

But since

$$\rho_0 u_0 \delta = \int_0^\delta \rho_0 u_0 \, dy$$

$$\delta^* = \int_0^\delta \left(1 - \frac{u}{u_0}\right) dy$$

The displacement thickness δ^* can be considered as a measure of the mass flow deficit due to the boundary layer velocity profile which could be written directly, for unit width, as $\int_0^\delta \rho_0 (u_0 - u) dy$. Equating this to the mass flow which would have occurred between the solid boundary and the displacement surface for unit width $\rho_0 u_0 \delta^*$

$$\rho_0 u_0 \delta^* = \int_0^\delta \rho_0 (u_0 - u) dy$$

$$\delta^* = \int_0^\delta \left(1 - \frac{u}{u_0}\right) dy$$

If δ^* is very small compared with δ then the velocity profile is very full and the boundary layer has a minimum effect on the displacement of the external flow. If δ^* approaches $\delta/2$ then the external flow is restricted by about half the boundary layer thickness. In ducted flows this displacement causes fluid in the centre of the duct to accelerate, while for external flows this displacement causes streamlines to move away from the body surface.

The *momentum thickness* θ is defined as the amount by which an inviscid flow would have to be displaced from the solid boundary to make the momentum of the resulting ideal flow equivalent to the momentum of the real flow through the boundary layer and the free stream. As the velocities in the boundary layer are less than that in the free stream, the boundary layer produces both a mass flow deficit and a momentum deficit. At any point in the boundary layer

momentum deficit = local mass flow deficit × local velocity

Integrating across the boundary layer

$$\text{momentum deficit} = \int_0^\delta \rho_0 (u_0 - u) u \, dy$$

This will be equal to the momentum of the free stream fluid which would pass through the space between the momentum surface and the solid boundary so that for unit width

$$\rho_0 u_0^2 \theta = \int_0^\delta \rho_0 (u_0 - u) u \, dy$$

$$\theta = \int_0^\delta \frac{u}{u_0} \left(1 - \frac{u}{u_0}\right) dy \tag{1}$$

If θ is small compared with δ, the boundary layer velocity profile is full and the momentum deficit low.

The rate of growth of the momentum thickness in the x direction $d\theta/dx$ is proportional to the local shear force between the fluid and the surface.

(b) Since

$$\delta^* = \int_0^\delta \left(1 - \frac{u}{u_0}\right)dy \quad \text{and} \quad \frac{u}{u_0} = 2\frac{y}{\delta} - \frac{y^2}{\delta^2}$$

$$\delta^* = \int_0^\delta \left(1 - 2\frac{y}{\delta} + \frac{y^2}{\delta^2}\right)dy$$

Putting

$$y/\delta = \eta \quad dy = \delta d\eta$$

The limits of integration become

$$y = 0, \quad \eta = 0$$
$$y = \delta, \quad \eta = 1$$

so

$$\delta^* = \delta \int_0^1 (1 - 2\eta + \eta^2)d\eta$$

$$= \delta \left[\eta - \eta^2 + \frac{\eta^3}{3}\right]_0^1 = \frac{1}{3}\delta$$

$$\frac{\delta^*}{\delta} = \frac{1}{3}$$

(c) Since

$$\theta = \int_0^\delta \frac{u}{u_0}\left(1 - \frac{u}{u_0}\right)dy \quad \text{and} \quad \frac{u}{u_0} = \left(\frac{y}{\delta}\right)^{1/7}$$

$$\theta = \int_0^\delta \left\{\left(\frac{y}{\delta}\right)^{1/7} - \left(\frac{y}{\delta}\right)^{2/7}\right\}dy$$

Putting

$$\frac{y}{\delta} = \eta$$

$$\theta = \delta \int_0^1 (\eta^{1/7} - \eta^{2/7})d\eta = \frac{7}{72}\delta$$

$$\frac{\theta}{\delta} = \frac{7}{72}$$

14.3 Momentum integral equation for boundary layer on a flat plate. General equation for boundary layer thickness δ

(a) Derive the momentum integral equation for boundary layer flow on a flat plate.

(b) Obtain a general equation for the boundary layer thickness δ for a laminar boundary layer growing on a smooth flat plate. Assume that the velocity profile in the boundary layer is

$$\frac{u}{u_0} = 2\frac{y}{\delta} - \frac{y^2}{\delta^2}$$

Solution (a) For a boundary layer growing on a smooth flat plate the streamlines are practically parallel to the surface so that the static pressure is everywhere the same. The shear forces are therefore the only forces acting on the fluid for both laminar and turbulent boundary layers.

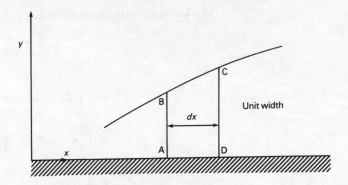

Figure 14.4

Considering a control volume ABCD, Fig. 14.4, of unit thickness perpendicular to the diagram at a distance x from some origin and of thickness dx, the rate of change of momentum of the fluid in the boundary layer as it passes through this control volume must equal the shear force exerted by the solid boundary on the fluid.

Fluid flows steadily into the control volume across AB and BC and leaves across CD.

Mass per unit time crossing AB $= \int_0^\delta \rho_0 u\, dy$ per unit width
so that

Momentum per unit time crossing AB $= \int_0^\delta \rho_0 u^2\, dy$ per unit width

Again

Mass per unit time entering through BC from the free stream

$$= \frac{d}{dx}\left\{ \int_0^\delta \rho_0 u\, dy \right\} dx \text{ per unit width}$$

and since all mass crossing BC possesses the free stream velocity u_0

Momentum per unit time crossing BC

$$= \frac{d}{dx} \left\{ \int_0^\delta \rho_0 u \, u_0 dy \right\} dx \text{ per unit width}$$

Also

Mass per unit time crossing CD

$$= \int_0^\delta \rho_0 u dy + \frac{d}{dx} \left\{ \int_0^\delta \rho_0 u dy \right\} dx \text{ per unit width}$$

Momentum per unit time crossing CD

$$= \int_0^\delta \rho_0 u^2 dy + \frac{d}{dx} \left\{ \int_0^\delta \rho_0 u^2 dy \right\} dx \text{ per unit width}$$

For steady flow in unit time

Change of momentum of fluid passing through ABCD

$$= \text{momentum crossing AB} + \text{momentum crossing BC}$$
$$- \text{momentum crossing CD}$$

$$= \int_0^\delta \rho_0 u^2 dy + \frac{d}{dx} \left\{ \int_0^\delta \rho_0 u \, u_0 dy \right\} dx -$$

$$\int_0^\delta \rho_0 u^2 dy - \frac{d}{dx} \left\{ \int_0^\delta \rho_0 u^2 dy \right\} dx$$

$$= \frac{d}{dx} \left\{ \int_0^\delta \rho_0 (u^2 - u \, u_0) dy \right\} dx$$

This will be equal to the shear force acting along AD $= -\tau_w dx$, the negative sign indicating that the shear force acts to oppose motion thus

$$\tau_w = \frac{d}{dx} \int_0^\delta \rho_0 u (u_0 - u) dy = \rho_0 u_0^2 \frac{d}{dx} \int_0^\delta \frac{u}{u_0} \left(1 - \frac{u}{u_0} \right) dy$$

But the momentum thickness

$$\theta = \int_0^\delta \frac{u}{u_0} \left(1 - \frac{u}{u_0} \right) dy$$

and so

$$\tau_w = \rho_0 u_0^2 \frac{d\theta}{dx} \tag{1}$$

Defining the local skin friction coefficient as $C_f = \tau_w / \frac{1}{2} \rho_0 u_0^2$ the momentum integral equation can be written

$$\frac{C_f}{2} = \frac{d\theta}{dx}$$

(b) If $\dfrac{u}{u_0} = 2\dfrac{y}{\delta} - \dfrac{y^2}{\delta^2}$ then from Example 14.2(a)

$$\theta = \int_0^\delta \left\{ 2\frac{y}{\delta} - \frac{y^2}{\delta^2}\left(1 - \frac{2y}{\delta} + \frac{y^2}{\delta^2}\right)\right\} dy$$

Changing the variable from y to η where $\eta = y/\delta$ when $y = 0$, $\eta = 0$ and when $y = \delta$, $\eta = 1$ so that

$$\theta = \delta \int_0^1 \{2\eta - \eta^2(1 - 2\eta + \eta^2)\} d\eta = \frac{2}{15}\delta$$

Substituting in the momentum integral equation (1) above

$$\tau_w = \rho_0 u_0^2 \frac{2}{15}\frac{d\delta}{dx}$$

In laminar flow the shear stresses are due to viscosity so that

$$\tau = \mu_0 \frac{du}{dy}$$

Since τ_w is the shear stress between the fluid and the solid surface or wall when $y = 0$,

$$\tau_w = \mu_0 \left(\frac{du}{dy}\right)_{y=0}$$

Now

$$u = u_0\left(2\frac{y}{\delta} - \frac{y^2}{\delta^2}\right) \quad \text{so that} \quad \left(\frac{du}{dy}\right) = u_0\left(\frac{2}{\delta} - \frac{2y}{\delta^2}\right)$$

giving

$$\left(\frac{du}{dy}\right)_{y=0} = \frac{2u_0}{\delta}$$

therefore

$$\tau_w = \mu_0 2\frac{u_0}{\delta} = \rho_0 u_0^2 \frac{2}{15}\frac{d\delta}{dx}$$

separating the variables

$$\delta.d\delta = 15\left(\frac{\mu_0}{\rho_0 u_0}\right) dx$$

Integrating $\dfrac{\delta^2}{2} = 15\left(\dfrac{\mu_0}{\rho_0 u_0}\right)x + C$ and for a boundary layer grow-

ing from the leading edge $\delta = 0$ at $x = 0$ so that $C = 0$ and

$$\delta = 5 \cdot 48\left(\frac{\mu_0}{\rho_0 u^0}\right)^{1/2} x^{1/2}$$

The quantity $\rho_0 u_0/\mu_0$ is known as the unit Reynolds number Re_u for the external flow and has dimensions L^{-1}, thus

$$\delta = 5 \cdot 48\, (Re_u)^{-1/2} x^{1/2}$$

14.4 Drag on a rectangular flat plate.

(a) Use the momentum integral equation to obtain a general expression for the drag coefficient C_D of a rectangular plate, span b and streamwise length l, wetted both sides and with laminar boundary layers having the approximate velocity profile $\dfrac{u}{u_0} = \dfrac{y}{\delta}$.

(b) Calculate the power absorbed in skin friction drag by a thin rectangular fin having a span of $0 \cdot 8$ m and a streamwise length of $0 \cdot 15$ m when moving through still water at 12 m/s. For water, density $\rho_0 = 1000$ kg/m³ and viscosity $\mu_0 = 0 \cdot 0015$ Ns/m².

Solution (a) From Example 14.3 equation (1)

$$\tau_w = \rho_0 u_0^2 \frac{d\theta}{dx}$$

and from Example 14.2 equation (1)

$$\theta = \int_0^\delta \frac{u}{u_0}\left(1 - \frac{u}{u_0}\right)dy$$

Putting

$$\frac{u}{u_0} = \frac{y}{\delta} = \eta$$

$$\theta = \delta \int_0^1 (\eta - \eta^2)d\eta = \frac{\delta}{6}$$

thus

$$\tau_w = \frac{\rho_0 u_0^2}{6} \frac{d\delta}{dx}$$

and since

$$\tau_w = \mu_0 \left(\frac{du}{dy}\right)_{y=0}, \quad \tau_w = \mu_c \frac{u_0}{\delta}$$

giving

$$\frac{\mu_0 u_0}{\delta} = \frac{\rho_0 u_0^2}{6} \frac{d\delta}{dx}$$

separating the variables

$$\delta.d\delta = 6\left(\frac{\mu_0}{\rho_0 u_0}\right)dx$$

Integrating

$$\frac{\delta^2}{2} = 6\left(\frac{\mu_0}{\rho_0 u_0}\right)x + C$$

If $\delta = 0$ at $x = 0$, then $C = 0$

$$\delta = 3 \cdot 46 \left(\frac{\mu_0}{\rho_0 u_0}\right)^{1/2} x^{1/2}$$

Substituting

$$\tau_w = \mu_0 \frac{u_0}{\delta} = \frac{\mu_0 u_0}{3 \cdot 46} \left(\frac{\rho_0 u_0}{\mu_0} \right)^{1/2} x^{1/2}$$

which, when plotted along the plate, will appear as in Fig. 14.5. The shear stress would be infinite at the leading edge and asymptotic to the x axis.

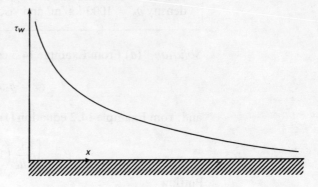

Figure 14.5

The total drag on a plate of span b and streamwise length l is obtained by integrating the elemental wall shear forces along the length of the plate. For a plate wetted on one side only

$$\text{Drag} = D = b \int_{x=0}^{x=l} \tau_w \, dx$$

For a plate wetted on both sides

$$D = 2b \int_0^l \tau_w \, dx$$

Substituting for τ_w

$$\text{Total drag on the plate} = \frac{2}{3 \cdot 46} \mu_0 u_0 \left(\frac{\rho_0 u_0}{\mu_0} \right)^{1/2} b \int_0^2 x^{-1/2} dx$$

$$= \frac{4}{3 \cdot 46} \mu_0 u_0 \left(\frac{\rho_0 u_0}{\mu_0} \right)^{1/2} bl^{1/2}$$

The *Drag coefficient* C_D for the plate is defined as

$$C_D = \frac{\text{Drag}}{\frac{1}{2} \rho_0 u_0^2 \times \text{wetted area}} = \frac{\text{Drag}}{\frac{1}{2} \rho_0 u_0^2 \times 2bl}$$

$$C_D = 1 \cdot 155 \, (Re_l)^{-1/2}$$

where Re_l is the flow Reynolds number based on the length of the plate in the streamwise direction.

(b) For a rectangular fin of the given dimensions

$$Re_l = \frac{1000 \times 12 \times 0 \cdot 15}{0 \cdot 0015} = 1 \cdot 2 \times 10^6$$

$$C_D = 1 \cdot 155 \left(\frac{1}{1 \cdot 2 \times 10^6}\right)^{1/2} = 1 \cdot 054 \times 10^{-3}$$

Total drag on fin $= C_D \times \tfrac{1}{2} \rho_0 u_0^2 \times 2lb$

$$= 1 \cdot 054 \times 10^{-3} \times 1000 \times 12^2 \times$$
$$0 \cdot 15 \times 0 \cdot 8 \text{ N}$$

$$= 18 \cdot 22 \text{ N}$$

Power absorbed $=$ drag force $\times$ velocity
$$= 18 \cdot 22 \times 12 = 218 \cdot 64 \text{ W}$$

14.5 Turbulent boundary layer on a smooth flat plate.

(a) Use the momentum integral equation to derive an equation relating boundary layer height δ with the distance x from the leading edge for a turbulent boundary layer on a smooth flat plate. Take the turbulent boundary layer velocity profile as

$$\frac{u}{u_0} = \left(\frac{y}{\delta}\right)^{1/7} \quad \text{and use the empirical equation for wall shear stress}$$

$$\tau_w = 0 \cdot 0225 \, \rho_0 u_0^2 \left(\frac{\mu_0}{\rho_0 u_0 \delta}\right)^{1/4}$$

Assume that the boundary layer starts from the leading edge.

(b) Derive an expression for the drag coefficient C_D for a rectangular plate of span b and streamwise length l wetted on one side only with a turbulent boundary layer having the same velocity profile and equation for wall shear stress as in (a). Assume that the turbulent boundary layer begins at the leading edge.

(c) An airship, 8 m diameter and 85 m long, travels through still air at 35 m/s. By treating the airship's cylindrical surface as if it were a rolled up flat plate wetted on one side only, calculate the power absorbed in skin friction.

Solution (a) The momentum integral equation derived in Example 14.3 equation (1) gives

$$\tau_w = \rho_0 u_0^2 \, \frac{d\theta}{dx}$$

and from Example 14.2 equation (1)

$$\theta = \int_0^\delta \frac{u}{u_0}\left(1 - \frac{u}{u_0}\right)dy \quad \text{or since} \quad \frac{u}{u_0} = \left(\frac{y}{\delta}\right)^{1/7}$$

$$\theta = \int_0^\delta \left\{\left(\frac{y}{\delta}\right)^{1/7} - \left(\frac{y}{\delta}\right)^{2/7}\right\} dy$$

Putting $\dfrac{y}{\delta} = \eta$

$$\theta = \delta \int_0^1 (\eta^{1/7} - \eta^{2/7})d\eta = \frac{7}{72}\delta$$

$$\frac{d\theta}{dx} = \frac{7}{72}\frac{d\delta}{dx}$$

and

$$\tau_w = 0 \cdot 0225\, \rho_0 u_0^2 \left(\frac{\mu_0}{\rho_0 u_0 \delta}\right)^{1/4} = \rho_0 u_0^2 \times \frac{7}{72}\frac{d\delta}{dx}$$

Separating the variables

$$\delta^{1/4}d\delta = 0 \cdot 231 \left(\frac{\mu_0}{\rho_0 u_0}\right)^{1/4} dx$$

Integrating

$$\frac{4}{5}\delta^{5/4} = 0 \cdot 231 \left(\frac{\mu_0}{\rho_0 u_0}\right)^{1/4} x + C$$

In most engineering applications boundary layers can be assumed to be either laminar or turbulent throughout. Occasionally it is necessary to proceed via a laminar boundary layer calculation up to some assumed transition point. The value of x and the height of the laminar boundary layer at transition are then used as the starting point, and to determine C for the above integral, for the subsequent turbulent boundary layer calculation.

Assuming the simple case of the boundary layer growing from the leading edge $\delta = 0$ when $x = 0$ giving $C = 0$ and

$$\delta = 0 \cdot 370 \left(\frac{\mu_0}{\rho_0 u_0}\right)^{1/5} \sigma^{4/5}$$

(b) If the plate is wetted on one side

$$\text{Drag coefficient } C_D = \frac{\text{drag force}}{\frac{1}{2}\rho_0 u_0^2 \times lb}$$

The drag force D will be the integral along the plate of the elemental drag forces $\tau_w\, b\, dx$

$$D = b \int_0^l \tau_w dx$$

From the question

$$\tau_w = 0 \cdot 0225\, \rho_0 u_0^2 \left(\frac{\mu_0}{\rho_0 u_0 \delta}\right)^{1/4}$$

and from part (a)

$$\delta = 0 \cdot 370 \left(\frac{\mu_0}{\rho_0 u_0}\right)^{1/5} x^{4/5}$$

$$\delta^{1/4} = 0 \cdot 78 \left(\frac{\mu_0}{\rho_0 u_0}\right)^{1/20} x^{1/5}$$

and so

$$\tau_w = 0 \cdot 0288 \, \rho_0 u_0^2 \left(\frac{\mu_0}{\rho_0 u_0} \right)^{1/5} x^{-1/5}$$

Substituting for τ_w in the equation for the drag D

$$D = b \int_{x=0}^{x=l} 0 \cdot 0288 \, \rho_0 u_0^2 \left(\frac{\mu_0}{\rho_0 u_0} \right)^{1/5} x^{-1/5} dx$$

$$= 0 \cdot 036 \, b \, \rho_0 u_0^2 \left(\frac{\mu_0}{\rho_0 u_0} \right)^{1/5} l^{4/5}$$

Drag coefficient $C_D = \dfrac{\text{drag}}{\frac{1}{2}\rho_0 u_0^2 lb}$

$$= 0 \cdot 072 \left(\frac{\mu_0}{\rho_0 u_0 l} \right)^{1/5}$$

$$= 0 \cdot 072 \, (Re_l)^{-1/5}$$

(c) From (b) the drag coefficient for a flat plate wetted on one side only is $C_D = 0 \cdot 072 \left(\dfrac{\mu_0}{\rho_0 u_0 l} \right)^{1/5}$

giving

$$C_D = 0 \cdot 072 \left(\frac{0 \cdot 000018}{1 \cdot 18 \times 35 \times 85} \right)^{1/5} = 0 \cdot 00158$$

The effective span of the rolled up plate $b = \pi d = \pi \times 8 = 25 \cdot 13$ m

Drag $D = C_D \times \frac{1}{2}\rho_0 u_0^2 \times lb = 0 \cdot 00158 \times 0 \cdot 5 \times 1 \cdot 18 \times 35^2 \times 85 \times 25 \cdot 13$

$$= 2439 \cdot 25 \text{ N}$$

Power absorbed = drag × velocity = $2439 \cdot 25 \times 35$ W
$$= 85 \cdot 373 \text{ kW}$$

Problems

1 A flat plate 1 m wide and 4 m long moves with a velocity of 4 m/s parallel to its longer sides through still air of kinematic viscosity $1 \cdot 5 \times 10^{-5}$ m²/s and density $1 \cdot 21$ kg/m³. On one side of the plate an initially laminar boundary layer is formed. On the other side the leading edge is roughened and the boundary layer may be considered to be entirely turbulent. Assuming a critical Reynolds number of 5×10^5 determine the ratio of the drag forces on the two sides.

For laminar flow $C_D = \dfrac{1 \cdot 46}{(Re_x)^{1/2}}$ and for turbulent flow

$$C_D = \frac{0 \cdot 074}{(Re_x)^{1/5}}$$

Answer $1 \cdot 5$ to 1

2 Air flows parallel to the surface of a smooth flat plate 10 m long. The boundary layer has zero thickness at the leading edge. The Reynolds number at the trailing edge of the plate is 10^7. Calculate the total drag force due to skin friction on one side of the plate per unit width.

Assume that for a laminar boundary layer, up to $Re_x = 5 \times 10^5$, the skin friction coefficient is $C_f = 1 \cdot 328 \, (Re_x)^{-1/2}$ and for a turbulent boundary layer $C_f = 0 \cdot 074 \, (Re_x)^{-1/5}$. Take the density of air as $1 \cdot 2 \text{ kg/m}^3$ and its dynamic viscosity as $1 \cdot 8 \times 10^{-5} \text{ kg/ms}$.

Answer $3 \cdot 75$ N

3 Calculate the thickness of a laminar boundary layer immediately prior to transition when water at 10° C flows over a smooth flat surface with a sharp leading edge under conditions of zero pressure gradient. The boundary layer velocity profile is described by

$$\frac{u}{u_0} = 3\left(\frac{y}{\delta}\right) - 2\left(\frac{y}{\delta}\right)^2.$$

The transition Reynolds number can be taken as 4×10^5 $u_0 = 5$ m/s and $(\mu_0)_{10°C} = 1 \cdot 308 \times 10^{-3} \text{ kg/ms}$

Answer $0 \cdot 906$ mm

4 Using the laminar flow boundary layer velocity equation

$$\frac{u}{u_0} = \sin\left(\frac{\pi}{2} \cdot \frac{y}{\delta}\right)$$

obtain a general expression for shear stress distribution on a flat plate and find the distance from the leading edge where the shear stress is equal to 10 Pa when the fluid is water of density $\rho_0 = 1000 \text{ kg/m}^3$ and viscosity $\mu_0 = 0 \cdot 0018 \text{ Ns/m}^2$ which is flowing at 15 m/s

Answer $2 \cdot 25$ mm

5 Two thin flat plate damper blades are parallel and aligned with the air flow in a duct carrying air at 25 m/s. Calculate the effective percentage reduction in flow area between the damper blades at their trailing edges if they are 200 mm apart, ignoring any side wall effects.

Assume laminar boundary layers throughout with a velocity profile

$$\frac{u}{u_0} = 2\frac{y}{\delta} - \left(\frac{y}{\delta}\right)^2$$

and that flat plate relationships can be used. For air density $\rho_0 = 1 \cdot 18$ kg/m³ and viscosity $\mu_0 = 1 \cdot 8 \times 10^{-5}$ Pa.s

Answer 0·638 per cent

6 Calculate the displacement thickness δ^*, the momentum thickness θ and the shape function $H = \delta^*/\theta$ for a laminar boundary layer, using the approximate velocity profile equation $U/u_0 = y/\delta$, at a distance of $x = 1$ m from the leading edge.

Obtain the expression for the slope of the displacement surface $d\delta^*/dx$ and find the rate at which the momentum thickness θ is increasing in the x direction. Take $u_0 = 10$ m/s, $\theta_0 = 0 \cdot 0013$ Ns/m², $\rho_0 = 1000$ kg/m³

Answer $\delta^* = 0 \cdot 624$ mm, $\theta = 0 \cdot 208$ mm, $H = 3$, $d\delta^*/dx = 3 \cdot 12 \times 10^{-4}$, $d\theta/dx = 1 \cdot 04 \times 10^{-4}$

7 A small rectangular fin attached to an aircraft has a span of 120 mm and a chord of 60 mm. Calculate the drag of this fin and the power absorbed when it moves through still air of density 0·9 kg/m³ and viscosity 0·000025 Ns/m² at a speed of 600 km/hr.

Assume that the boundary layer is laminar with a velocity profile

$$\frac{u}{u_0} = 3\left(\frac{y}{\delta}\right) - 2\left(\frac{y}{\delta}\right)^2$$

Answer 23 watts

8 During boundary layer flow studies on a high speed train, the gap between the carriages was found to cause the boundary layer thickness to increase suddenly by $\frac{1}{5}\delta_s$ where δ_s is the boundary layer thickness at the end of the carriage upstream of the gap. Calculate the boundary layer thickness at the end of the sixth carriage of a train moving at 180 km/hr through still air of density $\rho_0 = 1 \cdot 18$ kg/m³ and viscosity $\mu_0 = 0 \cdot 000019$ Ns/m². The train consists of a locomotive and six carriages each of which can be regarded as smooth slab-sided boxes 18 m long.

The boundary layer profile is $u/u_0 = (y/\delta)^{1/6}$ and the wall shear stress is given by

$$\tau_w = 0 \cdot 025\, \rho_0 u_0^2 \left(\frac{\mu_0}{\rho_0 u_0 \delta}\right)^{1/4}$$

Answer 1·089 m

9 An air intake on a flat roof is positioned facing directly into a wind of 15 m/s. The intake is square, 300 mm × 300 mm, with its bottom edge flush with a flat roof. A turbulent boundary layer grows for a distance of 40 m across the flat roof to the plane of the intake. Calculate the mass flow rate entering this air intake under zero pressure gradient conditions.

The boundary layer profile is

$$\frac{u}{u_0} = \left(\frac{y}{\delta}\right)^{1/7} \quad \text{and} \quad \tau_w = 0 \cdot 025\, \rho_0 u_0^2 \left(\frac{\mu_0}{\rho_0 u_0 \delta}\right)^{1/4}$$

Take $\rho_0 = 1 \cdot 18$ kg/m³ and $\mu_0 = 1 \cdot 3 \times 10^{-5}$ Pa.s

Answer 1·31 kg/s

10 Calculate the power absorbed in skin friction drag when an airship moves through still air at 80 km/hr. The airship can be assumed to be cylindrical, 5 m diameter and 35 m long. Air density $1 \cdot 15$ kg/m³, viscosity $0 \cdot 000016$ Ns/m².

$$\text{Turbulent boundary layer profile } \frac{u}{u_0} = \left(\frac{y}{\delta}\right)^{1/7}$$

$$\text{Wall shear stress } \tau_w = 0 \cdot 0225 \, \rho_0 u_0^2 \left(\frac{\mu_0}{\rho_0 u_0 \delta}\right)^{1/4}$$

Answer 7·04 kW

11 During experiments in which velocity profiles were measured at the trailing edge of a flat plate, the boundary layer profile above and below the plate was the expected 1/7th power law type with each boundary layer being 3 cm thick. If the air density was $1 \cdot 17$ kg/m³ and the undisturbed velocity was 31 m/s, calculate the drag force on the plate per metre span.

Take $\mu_0 = 1 \cdot 3 \times 10^{-5}$ Pa.s and $\tau_w = 0 \cdot 0225 \, \rho_0 u_0^2 \left(\frac{\mu_0}{\rho_0 u_0 \delta}\right)^{1/4}$

Answer 6·53 N

12 Oil of density 870 kg/m³ and viscosity $0 \cdot 006$ Ns/m² flows over a smooth flat surface. If the transition Reynolds number for this flow is 3×10^5, calculate the boundary layer thickness at a distance equal to twice the transition distance from the leading edge when the free stream velocity of the oil is 3 m/s.

$$\text{For the laminar boundary layer } \frac{u}{u_0} = 2\left(\frac{y}{\delta}\right) - \left(\frac{y}{\delta}\right)^2$$

$$\text{For the turbulent boundary layer } \frac{u}{u_0} = \left(\frac{y}{\delta}\right)^{1/7}$$

$$\text{Assume that the wall shear stress } \tau_w = 0 \cdot 0225 \, \rho_0 u_0^2 \left(\frac{\mu_0}{\rho_0 u_0 \delta}\right)^{1/4}$$

Answer $x_\tau = 0 \cdot 69$ m, $\delta = 26 \cdot 3$ mm

15

Flow under varying head

In all the problems of this chapter the head producing flow varies with time and therefore a solution can only be obtained by considering the state of affairs during a very small time interval δt and then integrating the relation obtained.

Time required to empty a reservoir

15.1 Tank emptying through orifice

Derive a formula for the time of emptying a vertical cylindrical tank through an orifice in the bottom. If such a tank is 1·8 m diam and the orifice in the bottom is 50 mm diam, find the initial height of the water above the orifice in order that 2·8 m³ of water will flow out in 395 s. Take C_d for the orifice as 0·6.

Solution. Referring to Fig. 15.1, at time t let the head producing flow be h, the rate of discharge through the orifice Q and the area of the free surface in the tank A. At time $t + \delta t$ suppose that the level has fallen δh as shown. For continuity of flow,

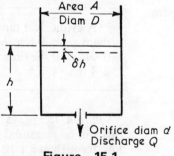

Figure 15.1

Change in contents of tank = Amount discharged through outlet

(Second contents − first contents)

= rate of discharge (second time − first time)

$$A(h - \delta h) - Ah = Q\{(t + \delta t) - t\}$$
$$-A\delta h = Q\delta t$$
$$\delta t = -\frac{A}{Q}\,\delta h \quad . \qquad . \qquad . \qquad . \quad (1)$$

Note. This equation applies to all emptying problems, whatever the means of emptying may be, and the time of emptying is found by integration after Q, and also A if it is not constant, have been expressed in terms of h.

For an orifice $Q = C_d a \sqrt{(2gh)}$ where a = orifice area. In this problem surface area A is constant. Substituting in equation (1)

$$\delta t = - \frac{A}{C_d a \sqrt{(2g)}} h^{-1/2} dh$$

If H_1 and H_2 are the initial and final values of h, then integrating

$$\text{Time of discharge} = T = \frac{2A}{C_d a \sqrt{(2g)}} (H_1^{1/2} - H_2^{1/2}) \qquad (2)$$

$$\text{Surface area } A = \tfrac{1}{4}\pi \times D^2 = \tfrac{1}{4}\pi \times (1{\cdot}8)^2 = 2{\cdot}55 \text{m}^2$$

If $2{\cdot}8$ m³ of water are discharged

$$\text{Fall in level } H_1 - H_2 = \frac{2{\cdot}8}{2{\cdot}55} = 1{\cdot}1 \text{m}$$

Also putting $T = 395$s, $C_d = 0{\cdot}6$ and $a = \tfrac{1}{4}\pi(0{\cdot}05)^2 \text{m}^2$. From equation (2):

$$H_1^{1/2} - H_2^{1/2} = \frac{395 \times 0{\cdot}6 \times \tfrac{1}{4}\pi(0{\cdot}05)^2 \sqrt{(2g)}}{2 \times 2{\cdot}55}$$

$$H_2^{1/2} = H_1^{1/2} - 0{\cdot}404$$

$$H_2 = H_1 - 0{\cdot}808 H_1^{1/2} + 0{\cdot}163$$

$$\text{Fall in level } H_1 - H_2 = H_1 - (H_1 - 0{\cdot}808 H_1^{1/2} + 0{\cdot}163) = 1{\cdot}1 \text{m}$$

$$0{\cdot}808 H_1^{1/2} = 1{\cdot}263$$

$$\text{Initial height} = H_1 = \mathbf{2{\cdot}45 m}$$

15.2 Tank emptying through pipe

A cylindrical tank, $0{\cdot}9$ m diam, is emptied through a 50 mm diam pipe $3{\cdot}6$ m long (both ends being sharp) for which $f = 0{\cdot}01$. Find the time taken for the head over the outlet to fall from $2{\cdot}4$ m to $1{\cdot}2$ m.

Solution. In Fig. 15.2, at any time t let the head producing flow be h. This will be measured from the free surface to the outlet. Considering flow through the pipe and applying Bernoulli's equation to the free surface and the outlet taking into account shock loss at entry

$$h = \frac{v^2}{2g} + \left\{ 0{\cdot}5 \frac{v^2}{2g} + \frac{4fL}{d} \cdot \frac{v^2}{2g} \right\}$$

$$v = \frac{\sqrt{(2g)}h^{1/2}}{\sqrt{\{1{\cdot}5 + (4fL)/d\}}}$$

Rate of discharge through pipe = $Q = \frac{1}{4}\pi d^2 v = \dfrac{\frac{1}{4}\pi d^2 \sqrt{(2g)}h^{1/2}}{\sqrt{\{1\cdot5 + (4fL)/d\}}}$

If level falls δh in time δt then, as shown in Example 15.1,

$$-A\delta h = Q\delta t$$

Taking A as constant and substituting for Q

$$\delta t = -\frac{A\sqrt{\{1\cdot5 + (4fL)/d\}}h^{-1/2}dh}{\frac{1}{4}\pi d^2\sqrt{(2g)}}$$

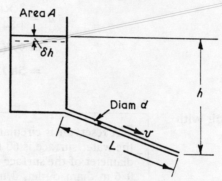

Figure 15.2

Integrating to find the time T for the level to fall from $h = H_1$ to $h = H_2$

$$T = -\frac{4A\sqrt{\{1\cdot5 + (4fL)/d\}}}{\pi d^2\sqrt{(2g)}}\int_{H_1}^{H_2} h^{-1/2}dh$$

$$= \frac{8A\sqrt{\{1\cdot5 + (4fL)/d\}}}{\pi d^2\sqrt{(2g)}}(H_1^{1/2} - H_2^{1/2})$$

Putting $A = \frac{1}{4}\pi \times (0\cdot9)^2\,\mathrm{m}^2$, $\quad d = 0\cdot05\,\mathrm{m}$, $\quad f = 0\cdot01$, $\quad L = 3\cdot6\,\mathrm{m}$, $H_1 = 2\cdot4\,\mathrm{m}$, $H_2 = 1\cdot2\,\mathrm{m}$

$$T = \frac{8 \times \pi \times 0\cdot81\sqrt{\{1\cdot5 + (4 \times 0\cdot01 \times 3\cdot6)/0\cdot05\}}}{4 \times \pi \times (0\cdot05)^2 \times \sqrt{(2g)}}(2\cdot4^{1/2} - 1\cdot2^{1/2})$$

$$T = \mathbf{138\cdot8\,s}$$

15.3 Reservoir emptying over weir

A reservoir has an area of 33 000 m² and discharges over a weir 3·6 m long which has its sill initially 0·6 m below the surface. Find the time required to reduce the level of the water in the reservoir by 0·5 m. Take the discharge of the weir as $Q = 1\cdot84Lh^{3/2}$ where h is the head measured above the sill.

Solution. If A = surface area of the reservoir and Q is the discharge over the weir under a head h at time t, then from Example 15.1,

$$-A\delta h = Q\delta t$$

Taking A as constant and substituting for Q in terms of h

$$\delta t = -\frac{A}{Q}\delta h = -\frac{A}{1\cdot84L}h^{-3/2}\delta h$$

Integrating from $h = H_1$ to $h = H_2$

$$\text{Time for level to fall} = T = \frac{2A}{1\cdot84L}\left(\frac{1}{H_2^{1/2}} - \frac{1}{H_1^{1/2}}\right)$$

Putting $A = 33000\,\text{m}^2$, $L = 3\cdot6\,\text{m}$, $H_1 = 0\cdot6\,\text{m}$, $H_2 = 0\cdot6 - 0\cdot5 = 0\cdot1\,\text{m}$

$$T = \frac{2\times33000}{1\cdot84\times3\cdot6}\left\{\frac{1}{(0\cdot1)^{1/2}} - \frac{1}{(0\cdot6)^{1/2}}\right\}$$

$$= 997(3\cdot162 - 1\cdot291) = 18680\,\text{s}$$

$$= \textbf{5h 11min 20s}$$

15.4 Reservoir with sloping sides

A reservoir is circular in plan and, when full, the diameter of the water surface is 60 m. When the water level falls $1\cdot2$ m the diameter of the surface is 48 m. Discharge takes place through a $0\cdot6$ m diam outlet, 3 m below high water level, which can be treated as an orifice with a coefficient of discharge of $0\cdot8$. Determine the time required to lower the water level $1\cdot2$ m if the reservoir is full at the start.

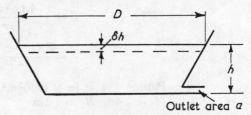

Figure 15.3

Solution. At time t let the head above the outlet be h (Fig. 14.3) and the diameter of the free surface D and suppose that in time δt the level falls δh. As shown in Example 15.1,

$$-A\delta h = Q\delta t$$

where Q = discharge through outlet.

Both Q and A vary with h and must be expressed in terms of h before integrating the equation to find the time of discharge.

Treating the outlet as an orifice, $Q = 0\cdot8a\sqrt{(2gh)}$. When $h = 3\,\text{m}$, $D = 60\,\text{m}$ and when $h = 1\cdot8\,\text{m}$, $D = 48\,\text{m}$. Thus $D = 30 + 10h$, and so

$$\delta t = -\frac{A}{Q}\delta h = \frac{-\frac14\pi(30 + 10h)^2}{0\cdot8\times\frac14\pi(0\cdot6)^2\sqrt{(2g)}}h^{-1/2}\delta h$$

$$= -\frac{(900 + 600h + 100h^2)h^{-1/2}\delta h}{1\cdot272}$$

$$dt = -(707h^{-1/2} + 471h^{1/2} + 78{\cdot}7h^{3/2})dh$$

Integrating from $H_1 = 3\,\text{m}$ to $H_2 = 3 - 1{\cdot}2 = 1{\cdot}8\,\text{m}$

Time for level to fall $= T = -\displaystyle\int_{3}^{1{\cdot}8} (707h^{-1/2} + 471h^{1/2} + 78{\cdot}7h^{3/2})dh$

$$= -\left[(1414h^{1/2} + 314h^{3/2} + 31{\cdot}5h^{5/2}) \right]_{3}^{1{\cdot}8}$$

$$= 1778\,\text{s} = \mathbf{29\,min\,38\,s}$$

Time required to fill a reservoir

15.5   Time to fill tank

> A circular tank, $1{\cdot}8$ m in diam and open to the atmosphere at the top, is supplied through a horizontal pipe 30 m long and 50 mm in diam entering the base of the tank. A pump feeds the pipe and maintains a constant gauge pressure of 45 kN/m² at the entry to the pipe. Find the time required to raise the level of water in the tank from $0{\cdot}9$ m to $1{\cdot}8$ m above the pipe inlet if $f = 0{\cdot}01$.

Solution. At time t suppose that the level in the tank is h above pipe inlet and the rate of inflow is Q.

Applying Bernoulli's equation to the free surface and to the pipe inlet (Fig. 15.4) and allowing for friction

$$H = h + \frac{fLQ^2}{3d^5}$$

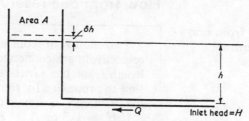

Figure   15.4

Inlet pressure $= 45\,\text{kN/m}^2 = 45000\,\text{N/m}^2$ gauge, and so

$$H = \frac{45000}{9{\cdot}81 \times 10^3} = 4{\cdot}58\,\text{m}$$

$f = 0{\cdot}01$, $L = 30\,\text{m}$, $d = 0{\cdot}05\,\text{m}$

$$4{\cdot}58 = h + \frac{0{\cdot}01 \times 30 \times Q^2}{3 \times (0{\cdot}05)^5}$$

$$Q = (14{\cdot}35 - 3{\cdot}125h)^{1/2} \times 10^{-3}$$

If the level in the tank rises δh in time δt

Second contents $-$ first contents $= Q$ (second time $-$ first time)

$$A\{(h + \delta h) - h\} = Q\{(t + \delta t) - t\}$$

$$\delta t = + \frac{Q}{A}\, \delta h$$

(Note the positive sign)

$$= \frac{A \delta h \times 10^3}{(14 \cdot 35 - 3 \cdot 125h)^{1/2}}$$

Put $(14 \cdot 35 - 3 \cdot 125h) = x$

then

$$\delta h = - \frac{\delta x}{3 \cdot 125}$$

and

$$\delta t = - \frac{A x^{-1/2} \delta x \times 10^3}{3 \cdot 125}$$

Integrating

$$T = - \frac{2A \times 10^3}{3 \cdot 125} \left[x^{1/2} \right]_{x_1}^{x_2}$$

where x_1 and x_2 are the values of x corresponding to h_1 and h_2.

Putting $A = \pi \times 0 \cdot 9^2\,\text{m}^2$, $h_1 = 0 \cdot 9\,\text{m}$, $h_2 = 1 \cdot 8\,\text{m}$

$$x_1 = 14 \cdot 35 - 3 \cdot 125 \times 0 \cdot 9 = 11 \cdot 54\,\text{m}$$

$$x_2 = 14 \cdot 35 - 3 \cdot 125 \times 1 \cdot 8 = 8 \cdot 73\,\text{m}$$

$$T = - \frac{2 \times \pi \times 0 \cdot 9^2 \times 10^3}{3 \cdot 125} (\sqrt{8 \cdot 73} - \sqrt{11 \cdot 54})$$

$$= \mathbf{721\,s}$$

Flow from one reservoir to another

15.6 Flow from one tank to another

> Two tanks of constant cross-sectional areas A_1 and A_2 respectively are connected by a pipe of diameter d which is of length L and has a friction coefficient f. Neglecting shock losses, find an expression for the time taken for the difference of level in the two tanks to change from H_1 to H_2.
>
> If $H_1 = 1 \cdot 8$ m, $A_1 = 8 \cdot 4$ m^2 and $A_2 = 4 \cdot 6$ m^2, find the time taken for $2 \cdot 8$ m^3 of water to pass from one tank to the other through a pipe 25 mm in diam and 150 m long if $f = 0 \cdot 01$.

Solution. The method is similar to that used for simple emptying but referring to Fig. 15.5 it can be seen that the change in the head producing flow is greater than the fall in level in the left-hand tank.

At time t let the difference of level producing flow be h and the velocity in the pipe v. Neglecting shock losses

$$h = \frac{4fL}{d} \frac{v^2}{2g}$$

$$v = \sqrt{\left[\frac{2gd}{4fL} \right]} h^{1/2}$$

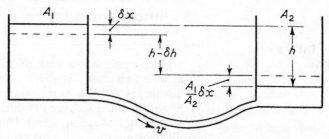

Figure 15.5

If the level in the left-hand tank falls δx in time δt,

$$\text{Amount passing from one tank to the other} = A_1 \delta x$$

$$\text{Rise in level in right-hand tank} = \frac{A_1}{A_2} \delta x$$

$$\text{Change in head producing flow} = \delta h = \delta x \left(1 + \frac{A_1}{A_2}\right)$$

Also,

Change in contents of left-hand tank

$$= \text{amount passing through pipe}$$
$$-A_1 \delta x = \tfrac{1}{4}\pi d^2 v \delta t$$

$$\delta t = -\frac{4A_1 \delta x}{\pi d^2 v}$$

Substituting for δx and v in terms of h

$$\delta t = -\frac{4A_1 A_2}{\pi d^2 (A_1 + A_2)} \sqrt{\left(\frac{4fL}{2gd}\right)} h^{-1/2} dh$$

Integrating from $h = H_1$ to $h = H_2$

$$T = \frac{8A_1 A_2}{\pi d^2 (A_1 + A_2)} \sqrt{\left(\frac{4fL}{2gd}\right)} (H_1^{1/2} - H_2^{1/2})$$

Note. This equation is symmetrical in A_1 and A_2, thus the time for a given change in the difference of level is the same irrespective of the direction of flow.

When $2 \cdot 8\,m^3$ of water leave left-hand tank

$$\text{Fall in level of l.-h. tank} = 2 \cdot 8 / 8 \cdot 4 = 0 \cdot 334\,m$$
$$\text{Rise in level of r.-h. tank} = 2 \cdot 8 / 4 \cdot 6 = 0 \cdot 609\,m$$
$$\text{Original difference of level} = H_1 = 1 \cdot 8\,m$$
$$\text{Final difference of level} = H_2 = 1 \cdot 8 - 0 \cdot 334 - 0 \cdot 609 = 0 \cdot 857\,m$$

Putting $A_1 = 8 \cdot 4\,m^2$, $A_2 = 4 \cdot 6\,m^2$, $d = 0 \cdot 025\,m$, $f = 0 \cdot 01$, $L = 150\,m$

$$T = \frac{8 \times 8 \cdot 4 \times 4 \cdot 6}{\pi (0 \cdot 025)^2 (8 \cdot 4 + 9 \cdot 6)} \sqrt{\left(\frac{4 \times 0 \cdot 01 \times 150}{2g \times 0 \cdot 025}\right)} (1 \cdot 8^{1/2} - 0 \cdot 857^{1/2})$$

$$= 12150 \times \sqrt{12 \cdot 25} \times (1 \cdot 3416 - 0 \cdot 9257)$$

$$= 17750\,s = \textbf{4h 55min 50s}$$

Inflow and outflow

15.7 Inflow and outflow

> A reservoir has a surface area of 80 000 m² which may be assumed constant. Water is entering at the rate of 9 m³/s and discharges to turbines at the rate of 4·25 m³/s. A spillway is also provided, the discharge of which can be taken as $28·6h^{3/2}$ m²/s where h is the head in metres. If the water level is originally below sill level how long will it take for the level to change from sill level to a head over the weir of 0·25 m?

Solution. At time t let the head over the spillway be h and suppose that in time δt the level rises δh. If $Q_1 = $ rate of inflow, $Q_2 = $ rate of discharge to turbines and $Q_3 = $ rate of discharge over spillway,

Increase in contents of reservoir

= inflow − discharge to turbines − waste over spillway

$$A\delta h = Q_1\delta t - Q_2\delta t - Q_3\delta t$$

$$\delta t = \frac{A\delta h}{Q_1 - Q_2 - Q_3}$$

Putting $A = 80000\,\text{m}^2$, $Q_1 = 9\,\text{m}^3/\text{s}$, $Q_2 = 4\text{·}25\,\text{m}^3/\text{s}$, $Q_3 = 28\text{·}6h^{3/2}$ m³/s

$$\delta t = \frac{80000}{4\text{·}75 - 28\text{·}6h^{3/2}} = \frac{2790}{0\text{·}167 - h^{3/2}}$$

In the limit

$$\frac{dt}{dh} = \frac{2790}{0\text{·}167 - h^{3/2}}$$

and

$$T = \int_{H_1}^{H_2} \frac{2790}{0\text{·}167 - h^{3/2}}\, dh$$

If dt/dh is plotted against h the area under the curve between $h = 0$ and $h = 0\text{·}25\,\text{m}$ is the required time T for the level to rise to $0\text{·}25\,\text{m}$ from sill level. Thus, the integration could be performed graphically or by using Simpson's rule as follows, using an increment δh of $0\text{·}025\,\text{m}$.

h m	0	0·025	0·050	0·075	0·100	0·125	0·150	0·175	0·200	0·225	0·250
$h^{3/2}$	0	0·004	0·011	0·021	0·032	0·044	0·058	0·073	0·089	0·107	0·125
$0\text{·}167 - h^{3/2}$	0·167	0·163	0·156	0·146	0·135	0·123	0·109	0·094	0·078	0·060	0·042
dt/dh	16700	17120	17900	19100	20650	22700	25500	29700	35800	46500	66400

$T = $ area under curve

$= \frac{1}{3}dh\{(\text{sum of first and last}) + 4 \times \text{even} + 2 \times \text{odd ordinates}\}$

$$= \frac{0 \cdot 025}{3} \{83\,100 + 4 \times 135\,120 + 2 \times 99\,850\}$$

$$= 6861\,\text{s} = \mathbf{1\,h\,54\,min\,21\,s}$$

Accelerating flow

15.8 Sudden opening of a valve

> A straight pipe, 360 m long, 100 mm in diam, with a sharp inlet, leads from a reservoir in which a constant level is maintained 15 m above the pipe outlet. The pipe outlet is initially closed by a valve, the frictional resistance of which may be taken as equivalent to 7·5 m length of pipe. If the valve is suddenly opened, estimate the time interval which must elapse before the velocity of flow in the pipe becomes 1·2 m/s. Neglect compressibility effects and take $f = 0 \cdot 08$.

Solution. At any time t let the velocity of flow in the pipe be v. Then if H is the constant head producing flow,

$H =$ friction head + velocity head at exit

$\qquad$ + acceleration head + loss at valve + shock loss

If L is the length and a the area of the pipe

$$\text{Mass of liquid in the pipe} = \frac{waL}{g}$$

If the velocity is $v + \delta v$ at time $t + \delta t$,

$$\text{Acceleration of liquid in pipe} = \frac{\delta v}{\delta t}$$

Thus

$$\text{Force required for acceleration} = \frac{waL}{g}\frac{dv}{dt}$$

$$\text{Head required for acceleration} = \frac{L}{g}\frac{dv}{dt}$$

$$\text{Head required to overcome friction} = \frac{4fL}{d}\frac{v^2}{2g}$$

$$\text{Loss of head at valve} = \frac{4f \times 7 \cdot 5}{d}\frac{v^2}{2g}$$

$$\text{Loss of head at entry} = 0 \cdot 5\frac{v^2}{2g}$$

$$\text{Velocity head at exit} = \frac{v^2}{2g}$$

Thus

$$H = \frac{v^2}{2g}\left(\frac{4f}{d}(L + 7\cdot5) + 1\cdot5\right) + \frac{L}{g}\frac{dv}{dt}$$

$$\frac{dv}{dt} = \frac{gH}{L} - \frac{cv^2}{2L}$$

where

$$c = \left(\frac{4f(L + 7\cdot5)}{d} + 1\cdot5\right)$$

$$dt = \frac{2L\,dv}{2gH - cv^2}$$

Integrating from $v = 0$ to $v = V$

$$\text{Time required} = T = \frac{2L}{c}\int_0^V \frac{dv}{\dfrac{(2gH)}{c} - v^2}$$

$$= \frac{2L}{c} \times \frac{1}{2 \times \sqrt{\left(\dfrac{2gH}{c}\right)}}\log_e \frac{\sqrt{\left(\dfrac{2gH}{c}\right)} + V}{\sqrt{\left(\dfrac{2gH}{c}\right)} - V}$$

$$= \frac{L}{\sqrt{(c \times 2gH)}}\log_e \frac{\sqrt{\left(\dfrac{2gH}{c}\right)} + V}{\sqrt{\left(\dfrac{2gH}{c}\right)} - V}$$

Putting $L = 360\text{m}$, $H = 15\text{m}$, $V = 1\cdot2\text{m/s}$, $d = 0\cdot01\text{m}$, $f = 0\cdot008$

$$c = \frac{4f}{d}(L + 7\cdot5) + 1\cdot5 = \frac{4 \times 0\cdot008 \times 367\cdot5}{0\cdot01} + 1\cdot5 = 119$$

$$T = \frac{360}{\sqrt{(119 \times 19\cdot62 \times 50)}}\log_e \frac{\sqrt{\{(19\cdot62 \times 15)/119\}} + 1\cdot2}{\sqrt{\{(19\cdot62 \times 15)/119\}} - 1\cdot2}$$

$$= 1\cdot925\log_e \frac{2\cdot775}{0\cdot375}$$

$$= 1\cdot95 \times 2\cdot00148 = 3\cdot9\text{s}$$

Problems

1 A vertical cylindrical tank, $0\cdot6$ m in diam and $1\cdot5$ m high, has an orifice of 25 mm diam in the bottom. The discharge coefficient is $0\cdot61$. If the tank is originally full of water, what time is required to lower the level by $0\cdot9$ m?
Answer   192 sec

2 A vertical cylindrical tank $1\cdot8$ m in diam has, at the bottom, a 50 mm diam sharp-edged orifice for which the coefficient of discharge is $0\cdot6$.
   (*a*) If water enters the tank at a constant rate of 9 dm³/s, find

the depth of water above the orifice when the level in the tank becomes steady. (b) Find the time for the level to fall from $2 \cdot 4$ m to $0 \cdot 6$ m above the orifice, if there is no inflow. (c) If water is now run into the tank at a constant rate of 17 dm³/s, the orifice remaining open, find the rate of rise in water level (cm per minute) when this level has reached $1 \cdot 5$ m above the orifice.
Answer $2 \cdot 97$ m; 775 s; 25 cm

3 Water is discharging from a bell-mouthed orifice ($C_d = 1 \cdot 0$) of 50 mm diam in the base of a tank having a surface area of 9 m². How long would it take to reduce the depth in the tank from $1 \cdot 2$ m to $0 \cdot 3$ m?
Answer 1135 s

4 A cylindrical vessel with its axis vertical is filled with water and discharges through an orifice 25 mm in diam at the bottom with a coefficient of discharge of $0 \cdot 623$. If the diameter of the vessel is $0 \cdot 6$ m, find the time required for the water level to drop from $1 \cdot 8$ m to $0 \cdot 6$ m above the orifice when the supply is cut off.
Answer 237 s

5 Discharge takes place from a 1 m diam cylindrical tank whose axis is vertical, through 3 m of 25 mm diam pipe. The pipe is connected to the base of the tank and discharges to atmosphere 2 m below the base. Initially the level in the tank is steady, water entering and leaving at a constant rate of $0 \cdot 002$ m³/s. If the supply of water is suddenly stopped, calculate the time required to empty it completely. Assume that the friction coefficient for the pipe is constant at $0 \cdot 01$ and that entry loss is $0 \cdot 5$ times the velocity head.
Answer 14 min 35 s

6 A pipe discharges water into a measuring tank $1 \cdot 8$ m long and $0 \cdot 9$ m wide. One wall of the tank has a V-notch through which the water escapes freely and a gauge is provided to measure the head of water above the bottom of the notch. When the pipe discharges steadily at the rate of 17 dm³/s the gauge shows 225 mm head. If the supply of water is suddenly cut off, how long will it be before the head falls to 75 mm?
Answer $59 \cdot 96$ s

7 Water flows over a sharp-edged rectangular notch $5 \cdot 4$ m wide, from a reservoir in which the water surface covers an area of 81 000 m². Find the time taken for the level to fall from $0 \cdot 9$ to $0 \cdot 15$ m above the sill of the notch, given that no water enters or leaves the reservoir by other means. Take C_d for the notch as $0 \cdot 62$, and assume that the reservoir sides are effectively vertical. Prove any formula used.
Answer 6 h 58 min 20 s

8 A sharp-edged V-notch inserted in the side of a rectangular tank $3 \cdot 6$ m long and $1 \cdot 2$ m broad gives a calibration $Q = 1 \cdot 44 H^{2 \cdot 5}$ where Q is measured in m³/s and H in m. Find how long it will take to reduce the head in the tank from $0 \cdot 3$ m to $0 \cdot 025$ m if the water discharges freely over the notch and there is no inflow into the tank.

Answer 8 min 14 s

9 A reservoir of water surface area 840 000 m², is provided with a spillway 30 m long which may be treated as a rectangular notch with coefficient of discharge 0·72. Find the time in hours for the head over the spillway to fall from 600 mm to 150 mm, neglecting inflow.

Answer 9·45 h

10 If the coefficient of discharge of a right-angled V-notch is 0·593, what will be the time required for flow through this notch to lower the surface level of a tank of 46 m² area from 0·45 m to 0·3 m above the vertex of the notch?

Answer 60·8 s

11 A vertical cylindrical tank is 4·8 m in diam and discharges through a pipe 90 m long and 225 mm in diam. How long will it take for the water level in the tank to fall from 2·7 m above the pipe exit to 1·2 m above that level? Assume $f = 0·01$.

Answer 460 s

12 A cylindrical tank 0·9 m in diam is placed with its axis vertical. The tank contains water and is emptied by a 25 mm diam pipe 2·4 m long, for which the coefficient of friction $f = 0·008$. The entrance to the pipe is smoothly rounded. The outer end of the pipe is 1·8 m below the bottom of the tank and is fitted with a nozzle of 12 mm diam with coefficient of discharge 0·98. Find (*a*) the volume of water discharged per sec when the water level in the tank is 1·2 m above the bottom, and (*b*) the time required to lower the water level from 1·2 m to 0·3 m above the bottom of the tank.

Answer 0·79 dm³/s, 13 min 9 s

13 A cylindrical tank 3 m in diam with its axis vertical is emptied through an open-ended 50 mm diam pipe which is 30 m long and has its far end 8·4 m below the connexion to the tank. The friction coefficient for the pipe is 0·006 and the loss at entrance may be taken as half the velocity head. How long will it take to drop the water level in the tank from 3·5 m to 0·6 m above the entrance to the pipe?

Answer 55 min 25 s

14 A rectangular tank 3·6 m × 1·2 m discharges through a 50 mm diam sloping pipe 3·6 m long with a valve. Find the time taken to empty the tank if the depth of water in the tank is originally 1·2 m and the outlet of the pipe is 0·9 m below the bottom of the tank. f for the pipe is 0·009 and the loss at the valve is 1·5 times the velocity head in the pipe. Entry to and exit from the pipe are both sharp.

Answer 19 min 35 s

15 The viscosity of a liquid is determined by timing the discharge of 50 cm³ of the liquid from a vessel through a horizontal capillary tube. The vessel is an upright cylinder open at the top, 5 cm in diam, and the capillary tube is of 1 mm bore and 10 cm long. The vessel is at first filled to a height of 5 cm above the axis of the tube and it is found that 50 cm³ are discharged in 20 min. Find the viscosity of the liquid in poises, given that the density is 0·88 g/cm³. Neglect the end effects of the tube and the velocity head at discharge.

Answer 0·0182 poise

16 A viscometer of the Redwood type has an oil-containing cylinder 4·75 cm in diam and an agate tube 0·17 cm in diam and 1·2 cm long. The oil surface, when flow starts, is 9 cm above the outlet from the agate tube. To allow for the sudden contractions at entry to the tube, the effective length of the tube may be taken as the actual length plus the tube radius. Making allowance for the decreasing head of oil, the viscous resistance through the tube, and the KE of discharge, calculate, using arithmetical integration, the time required for 50 cm³ of an oil of viscosity 0·5 poise and specific gravity 0·92 to flow through the viscometer.

Answer 244 s

17 The lower portion of a water tower reservoir is of conical form 1·8 m diam at the top reducing to 100 mm at the bottom in a depth of 0·9 m. A 100 mm diam pipe 1·8 m long projects vertically downwards from the bottom and discharges to atmosphere at its lower end. Calculate the time to empty this conical portion neglecting any friction or eddy losses except the pipe friction for which f may be taken as 0·01.

Answer 25 s

18 The diameter of an open-topped tank 1·5 m high increases uniformly from 4·2 m at the base to 6 m at the top. Discharge takes place through 3 m of a 75 mm diam pipe which opens to atmosphere 1·5 m below the base of the tank.

Initially the level in the tank is steady, water entering and leaving the tank at a constant rate of 17 dm³/s.

If the rate of inflow is suddenly doubled, find the time required to fill the tank completely.

Take f for the pipe as 0·01 and neglect only entry losses. Numerical or graphical integration will be accepted.

Answer 26 min 55 s

19 A cylindrical tank 2·4 m diam and 6 m long is placed with its axis horizontal. It may be emptied at the lowest point through a valve which may be considered as equivalent to a circular orifice 75 mm in diam with a coefficient of discharge of 0·6. If the valve is opened when the tank is full, calculate the time to reduce the level by 1·5 m assuming a free entry of air into the upper part of the tank.

Answer 20 min 55 s

20 A rectangular tank 3·6 m long and 1·8 m wide is divided into two parts by a vertical dividing plate so that the capacity of one portion is twice that of the other. The dividing plate has a hole 25 mm square near the bottom, forming a submerged orifice. At first the level of the water in the larger portion is 1·8 m above that in the smaller. Determine the time taken for this difference of level to be reduced to 0·9 m. Coefficient of discharge 0·6.

Answer 683 s

21 Two water tanks A and B, whose constant cross-sectional areas are 7·4 m² and 3·7 m² respectively, are connected by a 50 mm diam pipe, 120 m long, for which the friction coefficient f is 0·01. Given that the initial difference of level is 1·5 m and taking frictional resistance only into account, find the time taken for 2·25 m³ of water to pass from tank A into tank B.

Answer 42 min 25 s

22 Two cylindrical tanks, with axes vertical, stand on a horizontal floor. One is 1·8 m in diam, the other is 1·2 m in diam, and they are joined by a pipe, 75 mm diam and 1·8 m long, with sharp entrance and exit. The tanks are partly filled with water and at a given instant the level in the smaller tank is 1·2 m higher than the larger. Assuming f for the pipe is 0·009, calculate the time for the difference of level to become 0·3 m.

Answer 64·8 s

23 Two vertical-sided reservoirs each have a surface area of 186 m² and are connected by a submerged opening of area 0·186 m² which can be considered as an orifice with a coefficient of discharge of 0·8. If the initial difference of surface levels is 2·7 m, how long will it be before this difference is 1·2 m?

Answer 2 min 34·5 s

24 A tank 6 m long and 1·5 m wide is divided into two parts so that the area of one part is four times that of the other. The water level in the large portion is 3 m above that in the smaller. Find the time for the difference in water level in the two portions to reach 1·2 m if the water flows through an orifice in the partition 75 mm square for which $C_d = 0·6$.

Answer 123 s

25 Oil from a tank is continuously discharged by a pipe through which flow is laminar so that the rate of discharge can be expressed by $Q = kH$m³/s where Hm is the height of the oil level above the pipe outlet and k is a constant. The tank has straight sloping sides so that the area of the oil surface at height Hm above the pipe outlet is $(H^2 + 6H + 9)$m².

When oil enters the tank at a constant rate, the level in the tank remains constant so that $H = 1·2$ m. When the rate of inflow is doubled, H increases from 1·2 m to 1·8 m in 15 minutes.

If the inflow is completely cut off, find how long it will take for H to decrease from 1·8 m to its original value of 1·2 m above the pipe outlet.

Answer 514 s

26 A cylindrical tank 2·1 m diam and 2·4 m long is fixed with its axis vertical, and is open to the atmosphere. Water is fed in at the top at a constant rate.

Water flows from the bottom of the tank through a pipe 5·4 m long and 75 mm in diam, which discharges to atmosphere at a point 0·9 m below the bottom of the tank. When the water level is 0·6 m above the bottom of the tank it is found to rise at the rate of 25 mm in 19·5 s. Find the rate of inflow in dm³/s and determine whether the tank will overflow if this inflow rate is continued. Assume that the friction coefficient f is 0·008.

Answer 17·72 dm³/s; No

27 A reservoir with constant area of $2·3 \times 10^6$ m² has a spillway which discharges 10 m³/s when the head H is 225 mm. If water enters the reservoir from the catchment area at a constant rate of 17 m³/s, estimate the time taken for the head above the spillway to increase from 250 mm to 300 mm. Use arithmetic integration.

Answer 11 h 35 min 42 s

28 A hydraulic buffer stop has a cylinder 250 mm in diam and operates by forcing water through a 50 mm diam orifice for which the coefficient of discharge is 0·7. The discharge may be assumed to be against atmospheric pressure.

The buffer is used to arrest the motion of trucks, of mass 40 metric tons which strike it at 6 m/s. Ignoring the effect of friction, determine the time and distance travelled by the piston in reducing the velocity of the trucks to 20 per cent of the initial velocity.

Answer 2·06 m; 0·853 s

29 Two reservoirs whose constant difference of level is 12 m are connected by a pipe 100 mm in diam and 300 m long. Flow is controlled by a valve close to the lower reservoir and the loss of head across the valve when fully open is 10 $v^2/2g$, where v is the velocity in the pipe.

Assuming that the valve is suddenly opened and that the water is inelastic so that there is no pressure wave, find how long it takes for the velocity of flow in the pipe to attain a value equal to 0·95 of the steady terminal value.

Take $f = 0·01$ for the pipe and neglect any loss of head at pipe entry. Work from first principles or prove any formula used.

Answer 6·07 s

30 A hollow cone of height h and diameter d at the base is held with its axis vertical downwards, and is filled with water. A small circular hole whose diameter is $1/n$th that of the diameter of the base is made at the vertex. Assuming that the coefficient of discharge is equal to 0·62, show that the time taken for the depth of water to fall to one-half of its original value h cannot be less than

$$\frac{(4\sqrt{2} - 1)n^2\sqrt{h}}{12·4\sqrt{g}}$$

16

Uniform flow in open channels

An Open Channel

This is a duct through which a liquid flows with a free surface. At all points along its length the pressure on this free surface will be the same, usually atmospheric pressure. A channel may be covered, provided that it is not running full; a partly filled pipe is treated as an open channel. Since the pressure at the surface of the liquid is constant, flow is not due to pressure differences along the channel, but is caused by differences in the potential energy head due to the slope of the channel.

Uniform Flow

Under these conditions the cross-section of the flow is the same at all sections along the channel. The amount passing every section is the same and the mean velocity is constant. The frictional resistance to flow is then equal to the head due to the gradient of the bed of the channel.

Wetted Perimeter P

Taking a cross-section normal to the length of the channel, the wetted perimeter is the length of the line of contact between the liquid and the sides and base of the channel. For a rectangular channel of width B in which the depth of liquid is D,

$$P = B + 2D$$

Hydraulic Mean Depth m

If the area of the cross-section of the liquid is A,

$$m = \frac{A}{P}$$

For a rectangular channel $A = BD$ and

$$m = \frac{BD}{B + 2D}$$

Chezy formula

16.1 Chezy formula

> Assuming that the frictional resistance to flow in an open channel is proportional to the square of the mean velocity, derive the Chezy formula $v = C\sqrt{(mi)}$, in which v is the mean velocity of flow, m is the hydraulic mean depth, i is the bed slope of the channel, and C is a resistance coefficient. What are the dimensions of C?

Solution. Fig. 16.1 shows an element of fluid of length L and cross-sectional area A in the channel.

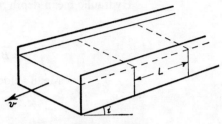

Figure   16.1

Force producing motion

= Component of weight of element acting along the channel

= $wAL \sin i = wALi$

if the slope is small.

If q is the frictional resistance per unit area at unit velocity,

Resistance per unit area at velocity $v = qv^2$

Frictional drag on element = $qv^2 \times$ area of contact with channel

= $qv^2 PL$

As the element is not accelerating,

Force producing motion = Frictional drag

$$wALi = qv^2 PL$$

$$v^2 = \frac{w}{q} \cdot \frac{A}{P} \cdot i = \frac{w}{q} \cdot m \cdot i$$

since $m = A/P$.

$$v = \sqrt{\frac{w}{q}} \; \sqrt{(mi)} = C\sqrt{(mi)}$$

To find the dimension of C,

$$C = \frac{v}{\sqrt{(mi)}}$$

and if v is measured in m/s, m in metres and i is an angle, C must be measured in $m^{1/2}/s$. Putting L for length and T for time, the dimensions of C are $L^{1/2}/T^{-1}$.

16.2   Rectangular section

A rectangular open channel has a width B of $4 \cdot 5$ m and a slope of 1 vertical to 800 horizontal. Find the mean velocity of flow v and the discharge Q when the depth D of water is $1 \cdot 2$ m, if C in the Chezy formula is 49 in SI units.

Solution. The Chezy formula is $v = C\sqrt{(mi)}$

Hydraulic mean depth $m = \dfrac{\text{area of flow}}{\text{wetted perimeter}}$

$$= \frac{BD}{B + 2D} = \frac{4\cdot5 \times 1\cdot2}{4\cdot5 + 2 \times 1\cdot2} = \frac{5\cdot4}{6\cdot9} = 0\cdot784\,\text{m}$$

Mean velocity $v = 49\sqrt{(0\cdot784 \times 1/800)}$

$$= 1\cdot53\,\text{m/s}$$

Discharge $Q = \text{area} \times \text{velocity} = BDV$

$$= 4\cdot5 \times 1\cdot2 \times 1\cdot53 = 8\cdot27\,\text{m}^3/\text{s}$$

16.3 V section

An open channel (Fig. 16.2) is V-shaped, each side being inclined at 45 deg to the vertical. If the rate of flow Q is $42\cdot5\ \text{dm}^3/\text{s}$ when the depth of water at the centre is 225 mm, calculate the slope of the channel using the Chezy formula, assuming that C is 49 in SI units.

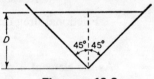

Figure 16.2

Solution. Using the Chezy formula,

$$Q = Av = AC\sqrt{(mi)}$$

Area of flow $A = D^2 = (0\cdot225)^2\,\text{m}^2$

Wetted perimeter $P = 2 \times D\sqrt{2}$

Hydraulic mean depth $m = \dfrac{A}{P} = \dfrac{D^2}{2\sqrt{2}D} = \dfrac{D}{2\sqrt{2}}$

$$Q = D^2C\sqrt{\left(\frac{D}{2\sqrt{2}} \times i\right)}$$

Squaring,

$$Q^2 = \frac{D^5C^2 i}{2\sqrt{2}}$$

and

$$i = \frac{2\sqrt{2}Q^2}{D^5C^2}$$

Putting $Q = 0\cdot0425\,\text{m}^3/\text{s}$, $D = 0\cdot225\,\text{m}$ and C = 49 SI units

Slope of the Channel $i = \dfrac{2\sqrt{2} \times (0\cdot0425)^2}{(0\cdot225)^5 \times 49^2} = \dfrac{1}{272}$

16.4 Trapezoidal section

The cross-section of an open channel is a trapezium with a bottom width B of $3\cdot6$ m and side slopes of 1 vertical to 2 horizontal. Assuming that C in the Chezy formula is 49 SI units, what will be the discharge Q if the depth of water D is $1\cdot2$ m and the slope i of the bed is 1 in 1600?

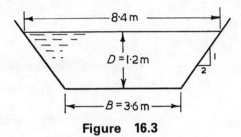

Figure 16.3

Solution. From Fig. 16.3

$$\text{Area of cross-section } A = \tfrac{1}{2}(8\cdot4 + 3\cdot6) \times 1\cdot2 = 7\cdot2\,\text{m}^2$$

$$\text{Wetted perimeter } P = B + 2\sqrt{(D^2 + 4D^2)}$$
$$= 3\cdot6 + 2\sqrt{(1\cdot44 + 5\cdot76)}$$
$$= 3\cdot6 + 5\cdot36 = 8\cdot96\,\text{m}$$

$$\text{Hydraulic mean depth } m = \frac{A}{P} = \frac{7\cdot2}{8\cdot96} = 0\cdot804\,\text{m}$$

$$\text{Discharge } Q = Av = AC\sqrt{(mi)}$$

$$= 7\cdot2 \times 49 \sqrt{\left(0\cdot804 \times \frac{1}{1600}\right)} = 7\cdot93\,\text{m}^3/\text{s}$$

16.5 Vertical walls with semicircular invert

A channel has vertical walls $1\cdot2$ m apart and a semicircular invert. If the centre-line depth is $0\cdot9$ m and the bed slope is 1 in 2500, what would be the value of C in the Chezy formula if the discharge is $0\cdot55$ m³/s?

Solution. From Fig. 16.4

$$\text{Area of cross-section } A = \tfrac{1}{2}(\pi \times 0\cdot6) + 1\cdot2 \times 0\cdot3 = 0\cdot926\,\text{m}^2$$

$$\text{Wetted perimeter } P = \pi \times 0\cdot6 + 2 \times 0\cdot3 = 2\cdot48\,\text{m}$$

$$\text{Hydraulic mean depth } m = \frac{A}{P} = \frac{0\cdot926}{2\cdot48} = 0\cdot372\,\text{m}$$

$$\text{Bed slope } i = \frac{1}{2500}, \quad Q = 0\cdot55\,\text{m}^3/\text{s}$$

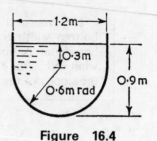

Figure 16.4

Using the Chezy formula,

$$Q = Av = AC\sqrt{(mi)}$$

$$0.55 = 0.926C\sqrt{\left(0.372 \times \frac{1}{2500}\right)}$$

$$C = \frac{0.55}{0.926 \times 1.187 \times 10^{-2}} = 50$$

16.6 Rectangular section

A rectangular channel has a width B of $2 \cdot 4$ m and a slope i of 1 in 400. What will be the depth of water D if the rate of flow Q is 85 m³/s and the coefficient C in the Chezy formula is 51 in SI units?

Solution. Using the Chezy formula, $v = C\sqrt{(mi)}$ where v is the mean velocity and m is the hydraulic mean depth.

$$\text{Rate of flow } Q = Av = AC\sqrt{(mi)}$$

$$\text{Cross-sectional area } A = BD$$

$$\text{Wetted perimeter } P = B + 2D$$

and

$$m = \frac{A}{P} = \frac{BD}{B + 2D}$$

from which

$$Q = BD \times C \times \sqrt{\left(\frac{BD}{B + 2D} \times i\right)}$$

$$8.5 = 2.4D \times 51 \times \sqrt{\left(\frac{2.4D}{2.4 + 2D} \times \frac{1}{400}\right)}$$

$$0.0694 = D\sqrt{\left(\frac{0.006D}{2.4 + 2D}\right)}$$

Squaring,

$$48.16 \times 10^{-4} = \frac{60 \times 10^{-4}D^3}{2.4 + 2D}$$

$$D^3 - 1.6D - 1.93 = 0$$

$$D = 1.66\text{m}$$

Other formulae

State the Manning formula for flow in channels, giving typical values of the roughness n.

A canal has a bottom width B of 3 m and sides with a slope of 1 vertical to 2 horizontal. The slope of the bed is 1 in 5000 and the depth of water D is $1 \cdot 2$ m. Using the Manning formula with $n = 0 \cdot 025$, calculate the rate of discharge in m^3/s.

Solution. By experiment it was found by Robert Manning that the Chezy coefficient C varied approximately as $m^{1/6}$. The Manning formula is

$$v = \frac{1}{n} m^{2/3} i^{1/2} \qquad \text{in SI units}$$

or

$$v = \frac{1 \cdot 486}{n} m^{2/3} i^{1/2} \qquad \text{in British units}$$

where n is a constant depending on surface roughness only, having the same value in both systems. Typical values are shown in the following Table.

Surface	n
Clean, smooth brick, stone or wood	0·010 to 0·017
Rubble masonry	0·017 to 0·030
Smooth earth	0·017 to 0·025
Rough earth	0·025 to 0·040
Irregular rock	0·035 to 0·045

$$\text{Top width} = B + 2 \times D \times 2 = 3 + 4 \cdot 8 = 7 \cdot 8 \, m$$

$$\text{Bottom width} = 3 \, m$$

$$\text{Cross-sectional area } A = \tfrac{1}{2}(3 + 7 \cdot 8) \times 1 \cdot 2 = 6 \cdot 48 \, m^2$$

$$\text{Wetted perimeter } P = B + 2\sqrt{\{(2D)^2 + D^2\}}$$

$$= 3 + 2\sqrt{(5 \times 1 \cdot 2^2)} = 3 + 5 \cdot 37 = 8 \cdot 37 \, m$$

$$\text{Hydraulic mean depth } m = \frac{A}{P} = \frac{6 \cdot 48}{8 \cdot 37} = 0 \cdot 775 \, m$$

Using the Manning formula

$$v = \frac{1}{n} m^{2/3} i^{1/2}$$

where $n = 0 \cdot 025$, $i = 1/5000$

$$v = \frac{1}{0 \cdot 025} (0 \cdot 775)^{2/3} \left(\frac{1}{5000} \right)^{1/2} = 0 \cdot 476 \, m/s$$

$$\text{Discharge } Q = Av = 6 \cdot 48 \times 0 \cdot 476 = \mathbf{3 \cdot 09 \, m^3/s}$$

16.8 Kutter formula and Bazin formula

(a) State the Kutter formula and the Bazin formula for the Chezy coefficient C.

(b) The friction coefficient k in Bazin's formula for an earth channel is $1 \cdot 3$. The channel is trapezoidal in cross-section with a bottom width of $1 \cdot 8$ m and side slopes of 1 vertical to 2 horizontal. Find the velocity of flow if the depth of water is $1 \cdot 5$ m and the slope is $0 \cdot 57$ m per km.

Solution. (a) Kutter's formula for C based on the analysis of the behaviour of rivers is

$$C = \frac{23 + \dfrac{0 \cdot 00155}{i} + \dfrac{1}{n}}{1 + \left(23 + \dfrac{0 \cdot 00155}{i}\right)\dfrac{n}{\sqrt{m}}}$$

where n has the same values as in Manning's formula, Example 16.7. Bazin's formula for C is

$$C = \frac{86 \cdot 9}{1 + k/\sqrt{m}} \text{ in SI units}$$

where k depends on the surface roughness.

(b) Bottom width of channel = $1 \cdot 8$ m

Top width = $1 \cdot 8 + 2 \times 2 \times 1 \cdot 5 = 7 \cdot 8$ m

Area of section = $A = \frac{1}{2}(1 \cdot 8 + 7 \cdot 8) \times 1 \cdot 5 = 7 \cdot 2$ m²

Wetted perimeter = $P = 1 \cdot 8 + 2\sqrt{(3^2 + 1 \cdot 5^2)} = 8 \cdot 5$ m

Hydraulic mean depth = $m = \dfrac{7 \cdot 2}{8 \cdot 5} = 0 \cdot 848$ m

Chezy coefficient C = $\dfrac{86 \cdot 9}{1 + (k/\sqrt{m})} = \dfrac{86 \cdot 9}{1 + (1 \cdot 3/\sqrt{0 \cdot 848})}$

$= 36$

Velocity of flow = $C\sqrt{(mi)}$

$= 36\sqrt{(0 \cdot 848 \times 0 \cdot 57 \times 10^{-3})}$

$= \mathbf{0 \cdot 784\,m/s}$

16.9 Darcy formula

Find an expression for the loss of head h_f in a channel of length L in terms of the velocity head and the hydraulic mean depth m.

Solution. From Example 16.1

$$v = C\sqrt{(mi)} \quad \text{or} \quad v^2 = C^2 mi$$

For uniform flow, $i = \text{bed slope} = \dfrac{h_f}{L}$

$$\text{Velocity head} = \frac{v^2}{2g} = \frac{C^2}{2g}\frac{mh_f}{L}$$

$$h_f = \frac{2g}{C^2} \times \frac{L}{m} \cdot \frac{v^2}{2g}$$

or putting $f = 2g/C^2$

$$h_f = \frac{fL}{m} \cdot \frac{v^2}{2g}$$

which is the Darcy formula, and is applicable to both pipes and channels. For a pipe, $m = \frac{1}{4}d$, giving

$$h_f = \frac{4fL}{d} \cdot \frac{v^2}{2g}$$

Proportions for maximum discharge or velocity

16.10 Proportions for maximum discharge

Find the proportions of a rectangular channel of depth D and width B which will make the discharge a maximum, for a given cross-sectional area A. Use the Chezy formula.

A canal is rectangular in cross-section and conveys $11\cdot3$ m³/s of water with a velocity of $1\cdot8$ m/s. Find the gradient required (a) if the proportions are those for maximum discharge, (b) if the width is three times the depth; C = 66 SI units.

Solution. For a given area A the channel could be either broad and shallow or deep and narrow or of any intermediate shape. Since A, C and the bed slope i are fixed and

$$\text{Discharge } Q = AC\sqrt{(mi)} = AC\sqrt{\left(\frac{A}{P}i\right)}$$

Q will be a maximum for the proportions which make P a minimum.

$$\text{Wetted perimeter } P = B + 2D$$

for a rectangular section.

$$\text{Area} = A = BD,$$

so that

$$B = \frac{A}{D}$$

$$P = AD^{-1} + 2D$$

Now P will be a minimum for a given value of A when

$$\frac{dP}{dD} = -AD^{-2} + 2 = 0$$

$$A = 2D^2$$

or, since
$$A = BD$$
$$BD = 2D^2$$

For maximum discharge
$$B = 2D$$

(a) If $B = 2D$, then
$$A = BD = 2D^2$$
$$Q = Av = 2D^2 v$$

Putting $Q = 11 \cdot 3\,\mathrm{m^3/s}$ and $v = 1 \cdot 8\,\mathrm{m/s}$
$$11 \cdot 3 = 3 \cdot 6 D^2$$
$$D = 1 \cdot 77\,\mathrm{m} \text{ and } B = 3 \cdot 54\,\mathrm{m}$$

Hydraulic mean depth $= m = \dfrac{A}{P} = \dfrac{2D^2}{B + 2D}$

$$= \frac{2D^2}{4D} = \frac{D}{2} = 0 \cdot 885\,\mathrm{m}$$

By the Chezy formula,
$$v = C\surd(mi)$$

$$\text{Gradient} = i = \frac{v^2}{C^2 m} = \frac{1 \cdot 8^2}{66^2 \times 0 \cdot 885} = \frac{1}{1190}$$

(b) If $B = 3D$, then
$$A = 3D^2$$
$$Q = 3D^2 v$$
$$11 \cdot 3 = 3 \times 1 \cdot 8 D^2$$

$$D = 1 \cdot 44\,\mathrm{m} \text{ and } B = 4 \cdot 32\,\mathrm{m}$$

Hydraulic mean depth $m = \dfrac{3D^2}{B + 2D}$

$$= \frac{3D^2}{5D} = \frac{3}{5} D = 0 \cdot 864\,\mathrm{m}$$

$$\text{Gradient} = i = \frac{v^2}{C^2 m} = \frac{1 \cdot 8^2}{66^2 \times 0 \cdot 864} = \frac{1}{1165}$$

16.11 Best trapezoidal section

Using the Chezy formula, find the proportions of a trapezoidal channel which will make the discharge a maximum for a given area. Show that the sides and base of such a section are tangential to a semi-circle whose centre is at the water surface.

Solution. Fig. 16.5 shows the cross-section.

$$\text{Area} = A = (B + nD)D$$
$$B = AD^{-1} - nD$$

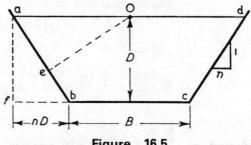

Figure 16.5

Wetted perimeter $P = B + 2\sqrt{(n^2 D^2 + D^2)}$
$$= B + 2D\sqrt{(n^2 + 1)}$$

Substituting for B

$$P = AD^{-1} - nD + 2D\sqrt{(n^2 + 1)}$$

Using the Chezy formula

$$Q = AC\sqrt{(mi)}$$

$$= AC\sqrt{\left(\frac{A}{P}i\right)}$$

For given values of A, C and i, Q will be a maximum when P is a minimum which occurs when

$$\frac{dP}{dD} = -\frac{A}{D^2} - n + 2\sqrt{(n^2 + 1)} = 0$$

$$\frac{A}{D^2} + n = 2\sqrt{(n^2 + 1)}$$

Putting $A = (B + nD)D$

$$\frac{B + nD}{D} + n = 2\sqrt{(n^2 + 1)}$$

Thus for maximum discharge the proportions are given by

$$\tfrac{1}{2}(B + 2nD) = D\sqrt{(n^2 + 1)}$$

To show that the sides and base are tangents to a semicircle, draw Oe perpendicular to ab from the centre of the water surface

$$\sin \widehat{Oae} = \frac{Oe}{Oa} = \frac{Oe}{\tfrac{1}{2}(B + 2nD)}$$

$$\sin \widehat{abf} = \frac{af}{ab} = \frac{D}{D\sqrt{(n^2 + 1)}}$$

Now $$\widehat{Oae} = \widehat{abf}$$

so that $$\frac{Oe}{\tfrac{1}{2}(B + 2nD)} = \frac{D}{D\sqrt{(n^2 + 1)}}$$

But for maximum discharge,

$$\tfrac{1}{2}(B \times 2nD) = D\sqrt{(n^2 + 1)}$$

and so
$$Oe = D$$

As ab is at right angles to Oe, it would be a tangent to a semicircle O and bc and cd are also tangents to this circle.

16.12 Best trapezoidal section

A trapezoidal channel to carry 142 m³/min of water is designed to have a minimum cross-section. Find the bottom width B and depth D if the bed slope i is 1 in 1200, the sides slope at 45 deg and $C = 55$ SI units.

Solution. For the cross-section to be a minimum the channel must be working under maximum discharge conditions and, from Example 16.11,

$$\tfrac{1}{2}(B + 2nD) = D\sqrt{(n^2 + 1)}$$

For 45 deg slopes, $n = 1$

$$B + 2D = 2D\sqrt{2} = 2{\cdot}828\,D$$

$$B = 0{\cdot}828\,D$$

$$\text{Area} = A = (B + nD)D = 1{\cdot}828\,D^2$$

$$\text{Wetted perimeter} = P = B + 2\sqrt{2}D = 3{\cdot}656\,D$$

$$\text{Hydraulic mean depth} = m = \frac{A}{P} = \frac{1{\cdot}828\,D^2}{3{\cdot}656\,D} = \frac{D}{2}$$

$$\text{Discharge} = Q = AC\sqrt{(mi)}$$

Putting $Q = 142/60\,\text{m}^3/\text{s}$, $C = 55$, $i = 1/1200$

$$\frac{142}{60} = 1{\cdot}828\,D^2 \times 55 \sqrt{\left(\frac{D}{2} \times \frac{1}{1200}\right)}$$

$$D^{2{\cdot}5} = 1{\cdot}149$$

$$\text{Depth} = D = \mathbf{1{\cdot}06m}$$

$$\text{Bottom width} = B = 0{\cdot}828\,D = \mathbf{0{\cdot}878m}$$

16.13 Depth for maximum velocity in circular section

Show that for a circular culvert of diameter D the velocity of flow will be a maximum when the depth of flow h at the centre is $0{\cdot}81\,D$. Use the Chezy formula.

A sewer, diameter $D = 0{\cdot}6$ m, has a slope i of 1 in 200. What will be the maximum velocity of flow that can occur, and what is the discharge at this velocity? Take $C = 55$ SI units.

Solution. In previous problems the area A has been kept constant and the shape and proportions of the channel varied to find the condition for maximum discharge. This cannot be done in the present problem, since the shape and size of the channel are fixed. Any change of depth of flow in the culvert produces a change of area.

In Fig. 16.6 the liquid surface subtends an angle 2θ at the centre O.

For any depth h

$$\text{Area of flow} = A = \text{sector OSTU} - \text{triangle OSU}$$

$$= \tfrac{1}{2}r^2 \times 2\theta - r^2 \sin\theta \cos\theta$$

$$= r^2(\theta - \tfrac{1}{2}\sin 2\theta)$$

$$\text{Wetted perimeter} = P = 2r\theta$$

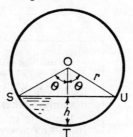

Figure 16.6

Using the Chezy formula

$$v = C\sqrt{\left(\frac{A}{P}i\right)}$$

and v will be a maximum for the value of θ which makes A/P a maximum

or

$$\frac{d(A/P)}{d\theta} = 0$$

$$\frac{d(A/P)}{d\theta} = \frac{1}{P^2}\left(P\frac{dA}{d\theta} - A\frac{dP}{d\theta}\right) = 0$$

$$P\frac{dA}{d\theta} = A\frac{dP}{d\theta}$$

Substituting for P, A, $(dA/d\theta)$ and $(dP/d\theta)$

$$2r\theta \times r^2(1 - \cos 2\theta) = r^2(\theta - \tfrac{1}{2}\sin 2\theta) \times 2r$$

$$\theta - \theta\cos 2\theta = \theta - \tfrac{1}{2}\sin 2\theta$$

$$2\theta = \tan 2\theta$$

giving

$$2\theta = 257\tfrac{1}{2}\,\text{deg}$$

$$\text{Depth of flow} = h = r - r\cos\theta$$

Putting $\theta = 128\tfrac{3}{4}\,\text{deg}$

$$h = r(1 + 0\cdot62) = 1\cdot62r$$

$$= 0\cdot81D$$

Putting $r = 0\cdot3\,\mathrm{m}$ and $2\theta = 257\tfrac{1}{2}\deg = 4\cdot5\,\mathrm{rad}$

$$A = 0\cdot3^2\left(\frac{4\cdot5}{2} + \frac{1}{2}\sin 77\tfrac{1}{2}°\right)$$

$$= 0\cdot09(2\cdot25 + 0\cdot976 \times \tfrac{1}{2}) = 0\cdot246\,\mathrm{m^2}$$

$$P = 2r\theta = 0\cdot6 \times 2\cdot25 = 1\cdot35\,\mathrm{m}$$

Thus Maximum velocity $v = \mathrm{C}\sqrt{\left(\dfrac{A}{P}i\right)}$

$$v = 55\sqrt{\left(\frac{0\cdot246}{1\cdot35} \times \frac{1}{200}\right)} = \mathbf{1\cdot61\,m/s}$$

$$\text{Discharge} = Av = 0\cdot246 \times 1\cdot61 = \mathbf{0\cdot397\,m^3/s}$$

16.14 Depth for maximum discharge in circular section

Assuming that the velocity of flow v in a circular channel of diameter D is given by $v = 66m^{1/2}i^{1/2}$ where m is the hydraulic mean depth and i is the bed slope, show that the discharge is a maximum when the depth of flow h at the centre is $0\cdot95\,D$.

Find the minimum diameter necessary if the channel has a slope of 1 in 5000 to give a discharge of $1\cdot4\,\mathrm{m^3/s}$.

Solution. Referring to Fig. 16.6, as in Example 16.13,

$$\text{Area} = A = r^2(\theta - \tfrac{1}{2}\sin 2\theta)$$

$$\text{Wetted perimeter} = P = 2r\theta$$

$$\text{Hydraulic mean depth} = m = \frac{A}{P}$$

$$\text{Discharge} = Q = Av = A \times 66m^{1/2}i^{1/2}$$

$$= 66\left(\frac{A^3}{P}\right)^{1/2} i^{1/2}$$

Since both A and P depend on θ, for maximum discharge A^3/P must be a maximum and

$$\frac{d(A^3/P)}{d\theta} = 0$$

or

$$\frac{1}{P^2}\left(3PA^2\frac{dA}{d\theta} - A^3\frac{dP}{d\theta}\right) = 0$$

$$3P\frac{dA}{d\theta} - A\frac{dP}{d\theta} = 0$$

Substituting in terms of r and θ

$$3 \times 2r\theta \times r^2(1 - \cos 2\theta) = r^2(\theta - \tfrac{1}{2}\sin 2\theta) \times 2r$$

$$4\theta - 6\theta\cos 2\theta = -\sin 2\theta$$

$$\theta = 154° = 2\cdot68\,\mathrm{rad}$$

Depth for max. discharge $= h = r(1 - \cos\theta)$

$$= \tfrac{1}{2}D(1 - \cos 154°) = \mathbf{0{\cdot}95D}$$

The minimum diameter is obtained under maximum discharge conditions, thus $\theta = 154°$.

$$A = r^2(\theta - \tfrac{1}{2}\sin 2\theta) = \tfrac{1}{4}D^2(2{\cdot}68 + 0{\cdot}394)$$
$$= 0{\cdot}77D^2$$

$$P = D\theta = 2{\cdot}68D$$

$$Q = 66\left(\frac{A^3}{P}\right)^{1/2} i^{1/2}$$

Putting $Q = 1{\cdot}4\,\text{m}^3/\text{s}$, $i = 1/5000$, and substituting for A and P

$$1{\cdot}4 = 66\left(\frac{(0{\cdot}77D^2)^3}{2{\cdot}68D}\right)^{1/2} \times \left(\frac{1}{5000}\right)^{1/2}$$

$$1{\cdot}96 = \frac{4356 \times 0{\cdot}456D^6}{2{\cdot}68D \times 5000} = 0{\cdot}1485$$

$$D^5 = 13{\cdot}2$$

$$D = \mathbf{1{\cdot}675\,m}$$

16.15 Maximum discharge by Manning formula

A $0{\cdot}9$ m diam pipe is to have a maximum discharge Q of $0{\cdot}7$ m^3/s, and it is found that the mean velocity is given by $v = 67i^{1/2}m^{2/3}$ where m is the hydraulic mean depth. Calculate the required value of the gradient i.

Solution. The condition for maximum discharge must first be found. The method is the same as in Example 16.14 but because v depends on $m^{2/3}$ instead of $m^{1/2}$ the solution is different.

$$Q = A \times v = A \times 67 \times i^{1/2}\left(\frac{A}{P}\right)^{2/3}$$

$$= 67 \times i^{1/2}\left(\frac{A^{2{\cdot}5}}{P}\right)^{2/3}$$

For maximum discharge $A^{2{\cdot}5}/P$ must be a maximum and the corresponding value of θ, Fig. 16.6, is given by

$$\frac{d(A^{2{\cdot}5}/P)}{d\theta} = 0$$

$$A = r^2(\theta - \tfrac{1}{2}\sin 2\theta)$$

$$P = 2r\theta$$

$$\frac{d(A^{2{\cdot}5}/P)}{d\theta} = \frac{1}{P^2}\left\{P \times 2{\cdot}5A^{1{\cdot}5}\frac{dA}{d\theta} - A^{2{\cdot}5}\frac{dP}{d\theta}\right\} = 0$$

Since $A^{1.5}/P^2$ is not zero,

$$2.5P\frac{dA}{d\theta} - A\frac{dP}{d\theta} = 0$$

Substituting in terms of r and θ

$$2.5 \times 2r\theta \times r^2(1 - \cos 2\theta) = r^2(\theta - \tfrac{1}{2}\sin 2\theta) \times 2r$$

$$3\theta - 5\theta\cos 2\theta + \sin 2\theta = 0$$

$$\theta = 2.64\,\text{rad} = 151\tfrac{1}{2}°$$

$$A = r^2(\theta - \tfrac{1}{2}\sin 2\theta)$$

$$= 0.45^2\left(2.64 + \frac{0.84}{2}\right) = 0.62\,\text{m}^2$$

$$P = 2r\theta = 2 \times 0.45 \times 2.64 = 2.37\,\text{m}$$

$$m = \frac{A}{P} = \frac{0.62}{2.37} = 0.262\,\text{m}$$

Discharge $Q = 67Ai^{1/2}m^{2/3}$

Putting $Q = 0.7\,\text{m}^3/\text{s}$

$$0.7 = 67 \times 0.62 \times i^{1/2} \times 0.262^{2/3}$$

$$i^{1/2} = \frac{0.7}{67 \times 0.62 \times 0.41} = 4.12 \times 10^{-2}$$

Bed slope $i = 16.97 \times 10^{-4} = $ **1 in 589**

Problems

1 A discharge of $47.5\,\text{dm}^3/\text{s}$ passes over a notch into a rectangular channel 0.375 m wide. Find the gradient of the bed of the channel for steady flow if the depth of water in the channel is 0.15 m. Take C = 66 SI units in $v = C\sqrt{mi}$.
Answer 1 in 510

2 Find the gradient necessary for a rectangular flume, 1.2 m wide and 0.6 m deep, to deliver $2.25\,\text{m}^3/\text{s}$ of water when running full. Use the Chezy formula and take C = 70 SI units.

If the flume is 8 km long and the water is used to drive turbines in a power station, what is the percentage loss in transmission if the difference of level between the upper end of the flume and the outlet from the turbines is 150 m?
Answer 1 in 151; 35.3 per cent

3 Working from first principles, derive the Chezy formula for uniform or normal flow in an open channel, and show the location of the hydraulic gradient line.

An open channel of rectangular section and breadth 0.3 m conveys water at a rate of $85\,\text{dm}^3/\text{s}$. C in the Chezy formula may

be taken as 58 and the slope as 2 in 1000. Estimate the depth for uniform flow.
Answer 0·338 m

4 A channel 0·9 m wide has vertical sides and the bottom is V-shaped, the angle of the V being 120 deg. If the depth of water flowing along the channel measured from the bottom of the V is 0·6 m, calculate the flow in m³/s when the bed has a slope of 1 in 1200.

Find the depth of water in the channel for the same flow when the V has silted up so that the bottom of the channel is now level and the slope of the bed remains unchanged. C = 55 SI units.
Answer 0·323 m³/s; 0·465 m

5 Assuming the resistance to flow in an open channel conveying water is proportional to the area of the wetted surface and to the square of the mean velocity v, show that $v = C\sqrt{mi}$ where m is the hydraulic mean depth, i the gradient of the bed and C a constant.

A channel of trapezoidal section with side slopes at 45 deg to the horizontal base of the channel, conveys water at a depth of 0·75 m. Find the width of the base and the gradient of the bed to discharge 1·27 m³/s with a mean velocity of 0·78 m/s. Take C = 66.
Answer 1·42 m, 0·000302

6 The section of an open channel is a semi-circular invert of 600 mm radius with vertical tangential sides. The slope of the channel is uniformly 1 in 2800. Find the flow in m³/s when the depth of water in mid-channel is 900 mm. Take C as 47 in SI units.
Answer 0·502 m³/s

7 A water channel is V-shaped with each side making an angle of 45 deg to the vertical. Calculate the volume of water passing per second when the depth of water in the channel is 0·25 m and the slope of the channel is 1 in 500. Take the coefficient C in the Chezy formula as 56 SI units. What would be the depth of water in the channel to pass twice this volume per second if the slope and value of C were unaltered?
Answer 46·4 dm³/s; 0·33 m

8 A rectangular open channel has a bottom width of 2·4 m and a surface roughness corresponding to $N = 0·015$ (Manning formula). If the slope of the bottom is 0·001 and the depth of flow 1·2 m, what is the discharge under conditions of uniform steady flow?

Compute the depth of flow in a channel of the same surface roughness but triangular in section with a 90-deg angle between the sides if the slope and quantity are to be the same as for the rectangular section above.
Answer 4·32 m³/s; 1·7 m

9 The cross-section of a conduit consists of vertical walls 1·2 m apart on a semi-circular invert. The slope is 1 in 1200. What

would be the flow when the centreline depth is 1 m, taking C in the Chezy formula as 72 in SI units.
Answer 1·34 m³/s

10 An irrigation channel has a gradient of 1 in 2000, a bottom width of 4·8 m and side slopes of 1 vertical to 2 horizontal. If the depth of water is 1·2 m and C = 50 SI units, what is the mean velocity and the discharge?
Answer 1·03 m/s; 8·9 m³/s

11 In a channel of rectangular cross-section with slope of 1 in 1000 the discharge is to be 1·4 m³/s when the channel, whose depth is half its breadth, is running full. If the channel is roughly excavated the value of C is 27 SI units. If the bottom and side slopes are smoothly finished C is 81. Assuming that the cost of excavation per cubic metre is twice the cost of smooth finishing per square meter compare the cost of the two types of channels.
Answer 1 to 0·98

12 Find the relation between the constant f and C in the alternative expressions for pipe flow $h_t = flv^2/2gm$ and $v = C\sqrt{(mi)}$.
 A circular brick sewer is 1·2 m in diam and has a fall of 1 in 500. Calculate the discharge of water when the depth of flow at the centre is 0·9 m. Take C = 52 in SI units.
Answer 1·28 m³/s

13 An open channel of trapezoidal cross-section cut in earth has a bottom width of 6 m, sides slopes 2 horizontal to 1 vertical, and bed slope 1 in 10 000. Determine the mean velocity v, and the discharge in m³/s for uniform flow when the depth of water at the centre is 2·4 m, using Manning's formula.

$$v = \frac{1}{N} m^{2/3} i^{1/2}$$

where N is the roughness coefficient = 0·025, m is the hydraulic mean depth, and i is the hydraulic gradient. Also find the corresponding value of C in the Chezy formula.
Answer 13·9 m/s; 43

14 Determine the maximum discharge from a circular stoneware sewer, 0·9 m diam, having a fall of 1 in 200; and compare your result with the discharge if the sewer is running full at atmospheric pressure. Assume maximum discharge when the wetted perimeter subtends an angle of 308 deg at the centre and take C = 55 in the Chezy formula.
Answer 1·235 m³/s; 1·175 m³/s

15 A trapezoidal channel is to be designed to convey 280 m³/min of water. Determine the cross-sectional dimensions of the channel if slope is 1 in 1600, side slopes 45 deg and the cross-section is to be a minimum. Take C = 50 in SI units.
Answer $D = 1·53$ m; $b = 1·27$ m

16 It is required to excavate a canal out of rock. It is to be of rectangular cross-section and to bring 14·2 m³ of water per second from a distance of 6·5 km with a velocity of 2·25 m/s. Determine the gradient and the most suitable section; C = 8·25 SI units.

Answer   1 in 1200; $D = 1·78$ m; $B = 3·56$ m

17   Determine the proportions of the most efficient open channel of trapezoidal section with sides inclined at 60 deg to the floor.

The flow in such a channel is found from

$$v = 58m^{0·57}i^{0·47}$$

Calculate the dimensions of a channel to discharge 7·5 m³/s if the slope is 1 in 1760.

Answer $B = (2/\sqrt{3})D$; 1·67 m; 1·93 m

18   Find the depth for maximum discharge and the depth for maximum velocity in a circular culvert 1·8 m in diam.

Answer   1·71 m; 1·46 m

19   The cross-section of a closed channel is a square with one diagonal vertical, s is the side of the square and y is the depth of the water line below the apex. Show that for the maximum discharge $y = 0·127s$ and that for maximum velocity $y = 0·414s$.

20   An irrigation channel of trapezoidal cross-section with side slopes at 60 deg to the horizontal is required to convey 4·25 m³/s when the slope of the bed is 1 in 9000. Determine suitable dimensions for the cross-section, which is to be a minimum, if C in the Chezy formula is 49 SI units.

Answer $D = 2·14$ m; $B = 2·47$ m

21   Determine the diameter of a circular conduit of slope i of 1 in 10 000 to give a maximum discharge of 2·8 m³/s, assuming that the velocity is given by $v = 80\ i^{1/2}m^{2/3}$ where m is the hydraulic mean depth.

Answer   2·4 m

22   Find an expression for the theoretical depth for maximum velocity in a closed circular channel in terms of the diameter d.

Compare the discharge at maximum velocity with that when the channel is running full, assuming that the Chezy formula is unaltered, and that the pressure remains atmospheric.

Answer   0·81 d; 0·964

23   An egg-shaped sewer has a section formed by circular arcs, the top being a semi-circle of radius R. The area and wetted perimeter of the section below the horizontal diameter of the semi-circle are $3R^2$ and $4·82R$ respectively. Prove that, if C in the Chezy formula is constant, the maximum flow will occur when the water surface subtends an angle of approximately 55 deg at the centre of curvature of the semi-circle.

24   The water supply for a turbine passes through a conduit which for convenience has its cross-section in the form of a square with one diagonal vertical. If the conduit is required to convey 8·5 m³/s, under conditions of maximum discharge, at

atmospheric pressure when the slope $i = 1$ in 4900, determine its size, assuming that the velocity of flow is given by $v = 80i^{1/2}m^{2/3}$.
Answer 2·99 m side

25 The upper portion of the cross-section of an open channel is a semi-circle of radius a; the lower portion is a semi-ellipse of width $2a$, depth $2a$, and perimeter $4·847a$, whose minor axis coincides with the horizontal diameter of the semi-circle. The channel is required to convey 14 m³/s when running three-quarters full (i.e. with three-quarters of the vertical axis of symmetry immersed), the slope of the bed being $i = 0·001$. Assuming the mean velocity of flow is given by Manning's formula $v = 80i^{1/2}m^{2/3}$, determine the dimensions of the section and the depth under maximum flow conditions.
Answer $a = 1·29$ m; 3·68 m

26 An open channel of "economic" trapezoidal cross-section with sides inclined at 60 deg to the horizontal is required to give a discharge of 10 m³/s when the slope of the bed is 1 in 1600. Calculate the dimensions of the cross-section, taking the constant M in Manning's formula ($v = Mi^{1/2}m^{2/3}$) as 74.
Answer $D = 1·82$ m; $B = 2·11$ m

27 Determine the ratio of the depth d to width b for a rectangular channel of constant area A and bed slope i for maximum discharge.

Determine the dimensions of such a channel given that $i = 1/2000$ and the discharge is 5·5 m³/s. What will be the discharge if the depth is reduced by 6 per cent? ($C = 66$).
Answer 1 to 2; 1·49, 2·97 m; 5·1 m³/s

Index